Field Guide to the larvae and British D

GW00467757

Volume 2: Damselflies (Zygoptera)

Steve Cham

The British Dragonfly Society

Published by The British Dragonfly Society

(Registered Charity No. 800196)

The British Dragonfly Society
23 Bowker Way, Whittlesey, Peterborough PE7 1PY

First published 2009

Credits
Text © 2009 Stephen Cham
Photographs © 2009 Stephen Cham
Typesetting and design by Stephen Cham

ISBN- 978-0-9556471-1-6

British Library Cataloging-in-Publication Data
A catalogue record for this book is available from the British Library.

To find out more about the British Dragonfly Society and where to send your records please visit the web site.
http://www.dragonflysoc.org.uk/

Front cover: Larva of Beautiful Demoiselle *Calopteryx virgo*
Frontispiece: Larva of Emerald Damselfly *Lestes sponsa*

Printed by Artisan Litho, Abingdon, Oxford.

Contents

Foreword   ii
Introduction   1
Anatomy of Larvae and Exuviae   2
The process of identification of larvae   10
Where to find Larvae   12
Where to find Exuviae   17
Emergence periods   21
A quick guide to size and shape   **22**

Species Accounts
Beautiful Demoiselle   *Calopteryx virgo*   24
Banded Demoiselle   *Calopteryx splendens*   27
Emerald Damselfly   *Lestes sponsa*   30
Willow Emerald Damselfly   *Lestes viridis*   32
Scarce Emerald Damselfly   *Lestes dryas*   33
Southern Emerald Damselfly   *Lestes barbarus*   35
White-legged Damselfly   *Platycnemis pennipes*   36
Large Red Damselfly   *Pyrrhosoma nymphula*   38
Small Red Damselfly   *Ceriagrion tenellum*   41
Southern Damselfly   *Coenagrion mercuriale*   44
Northern Damselfly   *Coenagrion hastulatum*   46
Azure Damselfly   *Coenagrion puella*   48
Caudal lamellae of Azure and Variable Damselflies - special notes   51
Variable Damselfly   *Coenagrion pulchellum*   52
A cautionary note when distinguishing Azure
and Variable Damselflies   57
Irish Damselfly   *Coenagrion lunulatum*   57
Common Blue Damselfly   *Enallagma cyathigerum*   58
Scarce Blue-tailed Damselfly   *Ischnura pumilio*   62
Blue-tailed Damselfly   *Ischnura elegans*   65
Red-eyed Damselfly   *Erythromma najas*   68
Small Red-eyed Damselfly   *Erythromma viridulum*   71

References / Suggested reading   74
Acknowledgements   75
Answer to ID challenge   75

Foreword

This second book, dedicated to damselflies, has been eagerly awaited by enthusiasts ever since publication of its companion volume on dragonfly larvae and exuviae back in 2007. Once again concise text and an array of superb photographs by Steve Cham lead you straight to the important identification features of each species without the need for tedious or complicated keys. But why is a guide like this really needed?

Most dragonfly enthusiasts reach this species group by way of a general love of natural history or a related interest in other flying creatures such as birds or butterflies. It is therefore the adults that first come to our notice. Later, as our interest in dragonflies develops, we realise the very special place they have in the environment. Due to their terrestrial needs and the fact that the major part of their life-cycle is aquatic, they are critical indicators of environmental quality both above and below the water.

However, just noting a damselfly or dragonfly's presence is not enough, we need to know where they are breeding at a site to show its conservation value. Only by identifying the larvae or exuviae that we find, can we know which areas are important for each species. Some prefer still or slow-moving water, whilst others breed in faster flowing streams and rivers. What they all have in common is their need for unpolluted water-bodies surrounded by suitable hunting and roosting terrain. As carnivores at or near the top of their respective food webs, they show the vigour of the ecosystem they are part of and the health of their habitats.

This guide will not only help you to identify the damselfly larvae or exuviae that you find, it will tell you the best times and places to look for them, explain their anatomy and lead you through the entire process from start to finish. I am incredibly proud to be President of a Society whose members can produce such a fantastic book.

Dr Pam Taylor, 2009
President of the British Dragonfly Society

Introduction

Dragonflies have become a popular group of insects in recent years, which have seen a plethora of field guides and books on the identification of adults. They have gained widespread appeal, especially amongst birdwatchers. Despite this, the study of the larval stages and exuviae remains a mystery to many. Their identification has still needed to be undertaken by the use of often complex keys, illustrated with line drawings of parts of the larval anatomy. This inherently restricts their appeal.

This guide to damselflies is the second volume to accompany that on dragonflies and is intended as an aid to the identification of larvae and exuviae in the field, without using identification keys. As such field workers, surveyors and site managers can identify British damselflies to genus or species whilst working on location. Using this guide no larvae need be harmed.

Damselflies are smaller and present more of a challenge for identification. The quick guide on pages 22-23 enables the reader to quickly compare the size and shape of a larva /exuvia against the life-size illustrations before moving on to the appropriate species account and detailed photographs.

PLEASE NOTE:
This field guide should only be used for larvae and exuviae of species of Damselfly that occur in Great Britain and Ireland. Some of the characteristics used to separate species may not be appropriate elsewhere. The characteristics described may only apply to final instar larvae unless otherwise stated.

This field guide focuses on characteristics that can be readily observed with the naked eye. Due to the small size of damselflies not all species can be identified by this alone. Therefore, a magnifying device, such as a 10x hand lens or stereo microscope with higher magnifications may be required. The three magnification symbols (below) are used throughout the field guide to indicate the best magnification to enable one to see the key diagnostic features.

 Eye symbol indicates that the feature is visible by the naked eye.

 Magnifier symbol indicates that the feature is best seen by using a 10x magnifier.

 Microscope symbol indicates that the feature is best seen by using higher magnification, such as a microscope.

Damselfly exuviae can prove difficult to identify in the field. The caudal lamellae often stick together which can make it difficult to see the key markings or features. It may be necessary to separate them before attempting identification. As the adults will be on the wing at the same time as the exuviae may be found it is advisable to look around for the former to help confirm identification. This may assist and narrow down the options of which species are present.

Larvae are best viewed from the side in a small water filled container. This enables the caudal lamellae to be clearly seen, making identification easier.

Anatomy of larvae and exuviae

Adults of all insect groups, including damselflies, have three main body parts: **Head, Thorax and Abdomen.** Damselflies undergo an incomplete metamorphosis. After the eggs hatch they go through a series of larval stages (instars) that are superficially similar to the adult. Each larval instar also exhibits the three main body parts. There can be 10-14 instars during larval development.

First impressions of body shape and size are good indicators of genus or species. Further detailed examination of distinguishing features should produce an identification of the species. Some of the features will require the use of magnification of 10x or higher. Hence, an understanding of the finer details of damselfly anatomy is needed to confirm identity.

> **Side view of larva of Red-eyed Damselfly** showing key body parts. Scale bar shows life size of body length and caudal lamellae.

Head

The head has prominent eyes which are widely separated. The top of the head should be examined as the shape and angles of the back of the head can be diagnostic. In some species this area can also have a distinctive spotted appearance.

The antennae are relatively small and inconspicuous in adult damselflies, but in larvae they play an important role in sensory perception. In some species they can be particularly long and serve as efficient mechano-receptors when moving around in plant debris. The antennae of Demoiselles (*Calopteryx* species) are especially long, with the first segment alone making up half of the total length.

The antennae tend to dry onto the head of exuviae and sometimes obscure parts of the eyes and mask (see next page). The number

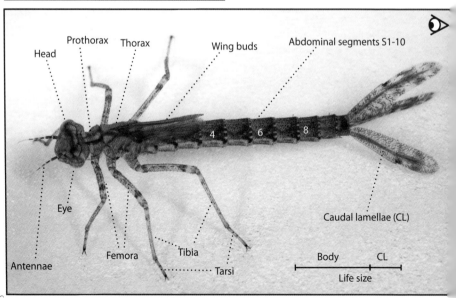

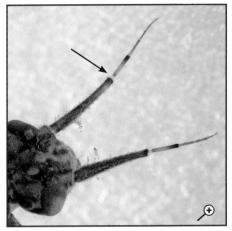

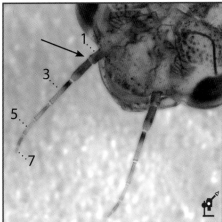

of segments can be useful to identify some species but may require a magnifier or microscope to see them.

The labial 'mask' is attached underneath the head and is the primary means of catching

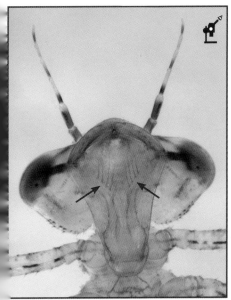

prey for larvae (see photograph overleaf). The mask lies flat under the head and persists through to the exuviae. The mask of Demoiselles and Emeralds have stemmed bases while all others have a basic elongated kite shape.

The mask comprises several parts :

The **prementum** forms the largest and most readily visible part of the mask and lies flat under the head when not in use. Its size and shape is important in the identification of a number of species.

The **labial palps** are attached to the front of the prementum. Depending on species these can bear movable hook or serrations, which enable them to catch and grasp prey whilst it is consumed.

Both the prementum and labial palps often bear hairs, or setae, the number of which varies between species.

To closely examine the setae the mask has to be removed. This is therefore not possible

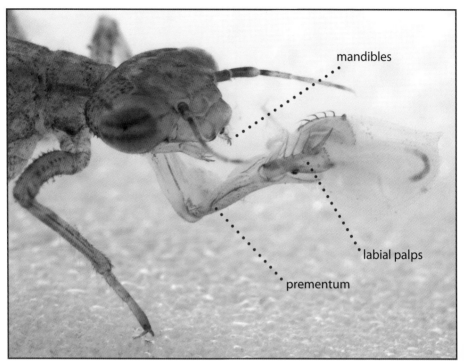

mandibles

labial palps

prementum

labial palps

prementum

Anatom

◀ **Larva of Azure Damselfly feeding on Water Flea** *Daphnia sp.*
The mask is rapidly projected out to catch the prey which is then held by the labial palps. The prey is pulled back to the mandibles. Note how the mask is articulated to the underside of the head. The labial palps hold the prey securely whilst it is consumed. The blue spheres in the lower photograph are unborn young of *Daphnia*.

with living larvae, but can be used with exuviae. For most British species of damselfly it is not necessary to use labial setae as a means of identification. They can be a useful diagnostic feature for confirmation, especially if the larva has lost its caudal lamellae.

The mandibles of larvae and exuviae are rarely visible, usually hidden by the mask. In larvae the mandibles can occasionally be seen whilst they are feeding. On exuviae the mask can be removed with a pair of fine tweezers, revealing the heavily pigmented mandibles below. *However, the mandibles have no use in identification*.

As the larva reaches the final instar it goes through metamorphosis in readiness for emergence as an adult. There is a progressive

expansion of the compound eyes at this stage. Although the eyes take up more of the front of the head this change should not affect identification.

Thorax

The wings and legs are attached to the thorax. Wing buds start to develop during the middle instars and increase in size throughout larval development. By the final instar the wing buds extend beyond the third abdominal segment (S3) onto the fourth (S4). Many key characteristics do not develop until later on in development and this means

Comparison of heads of final instar larvae of Azure Damselfly. The eyes change shape in the last instar and occupy more of the front of the head as the larva gets closer to emergence.

Anatomy

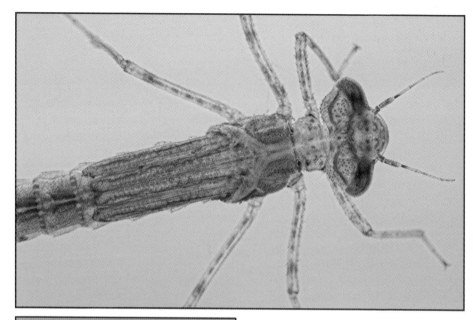

that the identification of earlier instars can be unreliable. For some early instars the body shape may enable identification to genus. Hence, **it is important when trying to identify larvae to species that final instar larvae are selected**.

It should be noted that in exuviae the wing buds are often displaced outwards as the adult damselfly emerges, leaving a large gap into the thoracic cavity.

The legs have three main parts; The femur is the first, long segment attached to the thorax. Then come the tibia and tarsi, the latter comprising several segments. Dark banding on the femora can be useful for identification of some species.
For some species the thorax bears markings that are distinctive enough to be used as reliable identification features.

Abdomen

The abdomen has ten segments (referred to as S1-10, S1 being the nearest to the head). S1 can be difficult to see, so it is best to count from S10. Both the dorsal (upper), ventral (under) and lateral (side) surfaces of the abdomen should be examined as they reveal features used in the identification of both larvae and exuviae. For example, Red -eyed Damselflies *Erythromma* spp have distinct rows of setae (coarse hairs) on the underside of S2 and S3 that are diagnostic.

The developing genitalia are located on the under surface of the abdomen. Female damselflies can be readily determined by

> ## CAUTION - Colouration
> **The colouration of exuviae and larvae is often highly variable and should not be used to determine identification.**
> Larvae can take on the colouration of the microhabitat in which they have lived, as well as being covered in debris.

Anatomy

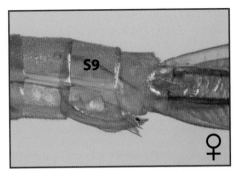

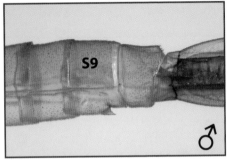

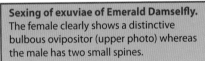

Sexing of exuviae of Emerald Damselfly.
The female clearly shows a distinctive bulbous ovipositor (upper photo) whereas the male has two small spines.

Underside view of exuviae of Small Red Damselfly.
This female clearly shows a distinctive bulbous ovipositor on S9.

the presence of a bulbous precursor to the ovipositor on S9 (see photos above). Similarly, the male can be determined by the lack of ovipositor and its replacement by a pair of small spines on S9. The location of the accessory genitalia of males can be seen as a slightly raised area on the middle of the under surface of S2 and S3. This will become the accessory genitalia in adult males.

Caudal lamellae

Damselfly larvae and exuviae possess three large and distinctive paddle-shaped caudal lamellae attached to the tip of the abdomen on S10. This distinguishes them from dragonfly larvae. These can take various forms and their relative size and shape can be important for the identification of species. In Demoiselles the outer lamellae are noticeably triangular in cross section. Caudal lamellae have two main tracheal trunks which each give off branches or secondary tracheae. These in turn give off tracheoles. In Emerald Damselflies the secondary branches are characteristically perpendicular to the main branch.

There are three forms of lamellae:-

Nodate lamellae- have a distinct node or notch at the mid position with a joint clearly visible across the lamellae. The margins of the caudal lamellae are fringed with fine setae, the size of which change abruptly at the node. The hairs between the node and the abdomen are called pre-nodal setae and

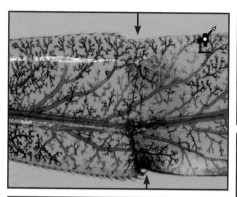

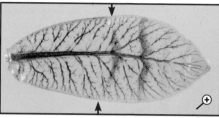

Middle caudal lamella of Red-eyed Damselfly showing distinct node at mid point (nodate).
Note the distinctive change in the size of the marginal hairs (setae) at this point (arrowed).

those from the node to the tip are post-nodal setae. The Red-eyed Damselfly is a good example of a species with nodate lamellae.

Subnodate lamellae- where there is no distinctive notch or joint, the position of the node is indicated by a change of size of the setae around the mid-point. This takes the form of a discontinuity or faint trace across the lamellae. This is sometimes offset from the mid point on one margin creating an obliquely nodate 'line'. The Common Blue Damselfly is a good example.

Denodate lamellae - no trace of a node or obvious difference in the setae. Large Red Damselfly is a good example.

It should also be noted that the caudal lamellae of early instar larvae do not exhibit the same characteristics as those of late and final instars, tending to be more elongate and pointed.

The caudal lamellae provide a useful character for the identification of most species of damselfly. This especially applies to species that have spots, bands and distinctive markings on the caudal lamellae. For larvae they can be examined by placing a larva in a small water filled observation chamber such as a small rectangular plastic pot, readily available from entomological suppliers. View side on. Species such as Large Red Damselfly exhibit a distinctive posture that greatly assists identification.

The dried, often distorted, lamellae of exuviae are best examined by removing them. Soak them in water with a few drops of vinegar or washing up liquid. This will soften them and enable a mounted needle to carefully prise them apart and away from the end of the abdomen. Lamellae can be examined by either making a temporary mount in water, drying them, or mounting them between a glass microscope slide and coverslip. The middle lamella is often the largest and with the most distinct outline.

Care should be taken when using the

Field tip:
When checking for markings on the caudal lamellae try holding specimens against a plain pale background such as the sky, or viewing them side on in a water filled container.

size and shape of the caudal lamellae for identification purposes. Occasionally, and without obvious effects on their well-being some or all three caudal lamellae can be lost. In some cases they may start to regrow giving them a shorter or distorted appearance. This can make it difficult to confirm an identification.

Larvae lose their caudal lamellae for various reasons. In captivity (and probably in the wild) larvae will often attack other larvae by striking at the caudal lamellae resulting in their loss. Caudal lamellae can sometimes fail to develop or grow abnormally as a result of damage incurred.

▼ Caudal lamellae of Azure Damselfly larva showing under-developed outer lamella.

▼ Caudal lamellae of Azure Damselfly larva showing abnormally developed middle and outer lamella.

Scarce Emerald Damselfly larva showing damaged and under-developed middle lamella.

Caudal lamellae of Emerald Damselfly larva showing under-developed outer lamella.

Caudal lamellae of Large Red Damselfly larva showing under-developed middle lamella.

The process of identification of larvae

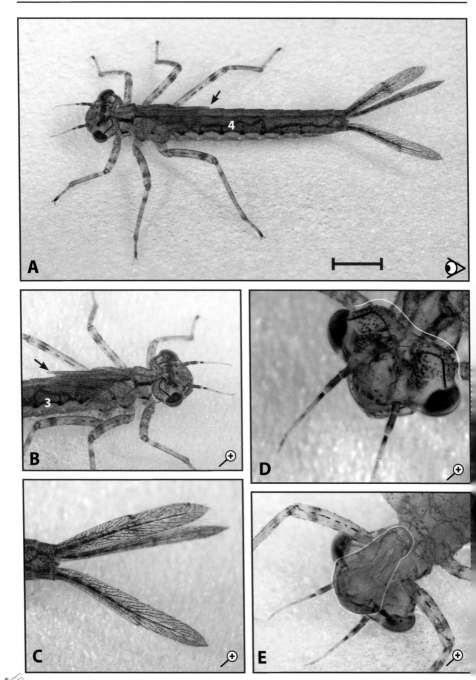

The process of identification is a combination of recognising and building up a series of key features or characteristics, before concluding the identification to species. There is also a process of elimination, where it may also be possible to say what it is not.

Size and shape

First impressions are important. The larva opposite is relatively small, being 14mm long including the caudal lamellae (**A**). There are no particularly distinctive markings on the body or caudal lamellae when viewed by the naked eye.

Thorax and abdomen

Looking closer with a magnifier it becomes apparent that the wing buds are relatively short, covering the first and second segments of the abdomen (**B**). It is clear that this is not a final instar larva and therefore identification to species level may be difficult. In a final instar larva the wing buds would project well over the fourth abdominal segment.

Caudal lamellae

Look at the caudal lamellae carefully with the magnifier (**C**) for markings. In their absence look for the presence or absence of a distinct

Damselfly larva of unknown identity
Figure A. Side view of exuvia. Bar indicates life size length of exuviae, including the caudal lamellae. Arrow indicates tip of wing buds.
Figure B. Top view of head and thorax. Arrow indicates tip of wing buds. S3 highlighted.
Figure C. Side view of caudal lamellae with distinct node.
Figure D. Head viewed from above (offset line drawn to show shape of hind margin).
Figure E. Head viewed from below. Shape of mask shown in outline

node or division between the thicker basal half and the rest of the lamella. Just by using a magnifier it is apparent that there is a node at the mid position of the caudal lamellae of this larva. This will immediately narrow down the options.

Head

The head should be examined with the magnifier, paying particular attention to the shape of the hind margin of the head (**D**). This might be curved or angular. On this larva the head's hind margin has a smooth curved shape rather than a more rectangular outline. Close inspection also reveals that the area at the back of the head has a number of bold spots. It also reveals that the antennae have 7 segments. At this stage the experienced observer would have a good idea of the genus if not species.
Examination of the underside of the head is also important; look at the mask. Demoiselles and Emeralds have distinctive masks with a narrow basal stem. This specimen, however, shows a kite-shaped mask which is typical of most damselflies (**E**).

Habitat

Finally consider the habitat in which the larva was netted. This can be of great help in eliminating some species. This specimen was taken from a small lake with abundant aquatic vegetation during late November and is therefore unlikely to be one associated with flowing waters.

Conclusion and ID

Is this enough information to make an identification? What might it be? Test yourself! Answer on page 75.

Where to find larvae

The duration of the life cycle of damselflies from egg to adult can vary from one year to two years, during which the majority of time is spent as a larva.

Age classes

A damselfly's life cycle starts when the female lays eggs by inserting them into plant material (endophytic oviposition). Some species, such as the Emerald Damselfly overwinter as eggs. Others overwinter as larvae and can be particularly difficult to find in the middle of winter when they move to the deeper, ice free water.

After hatching from the eggs, the larvae go through a series of instars during which they increase in size. From the time the eggs are laid there is considerable mortality in each age class. Therefore, more earlier instar larvae should normally be present than in each subsequent instar. However, in reality very few, if any, early instar larvae are found, often despite intensive searching. Early instar larvae (see photograph below) are so small that they are unlikely to be found in the net when pond dipping without very diligent work under a microscope.

When to look

Larvae utilise different types of aquatic micro-habitat at different times of year. Spring and Autumn are the best times to search for larvae. Spring is especially good as larvae are active and growing in readiness for emergence.

Larvae of some damselflies, such as the emerald damselflies *Lestes* spp., develop rapidly in the spring and early summer and therefore will be not encountered at other times of year. Larvae of some species can also be difficult to find during the summer months. This is due to the whole cohort of final instar larvae having recently emerged as adults. Only early instar larvae will be found at this time.

During the winter period it can be very difficult to find larvae, particularly at the margins and water's edge. There is evidence to suggest that larvae conceal themselves in deeper water to pass the colder winter months. Sampling in the margins in winter often fails to find any larvae where they would normally be present at other times.

Where to look

Damselfly larvae need to find micro-habitat conditions that provide them with enough cover or protection to avoid predation, as well as a plentiful food supply during the main growth periods. The most successful areas for larval sampling are those that offer cover

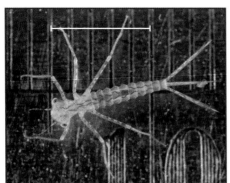

Early instar larvae- Larvae such as this, here photographed with a metric rule in the background are occasionally found in pond dipping samples after careful examination. Many go unnoticed. The scale bar is 5mm.

for the larvae as well as their prey. Such areas usually have sufficient underwater plants or organic debris to conceal the larvae. A good knowledge of the habitat requirements of adult dragonflies will enable one to predict the best places to search for larvae. All damselfly species tend to favour areas of habitat with submerged and marginal aquatic plants.

Some species can be elusive, especially if their preferred aquatic vegetation is out in the middle of a lake. Larvae of Small Red-eyed Damselflies have proved difficult to find due to their liking for Hornwort that often grows in open water.

What to use

One of the best ways to sample for damselfly larvae is to collect a large netful of plants and

Pond margin- a good place to pond dip for damselfly larvae. A microhabitat comprising a mixture of emergent, floating and submerged plants combined with rich organic debris on the bottom. Species include Large Red, Blue-tailed and Azure Damselflies.

spread them over a light coloured plastic sheet or tip them into a large white tray. The larger larvae will soon start to move and reveal themselves. However, early instar larvae below a certain size are rarely found by this method. When plant debris is removed

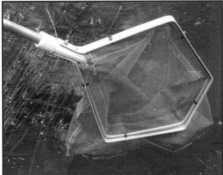

The specially designed 'larval' net used for sweeping through a range of different substrates.

Where to find larvae

from water the surface meniscus of water can easily trap small larvae and prevent them from moving. The weight of vegetation or debris removed from the water also hampers the movement of small larvae.

Larvae can be collected by using a variety of nets. When using nets on poles care should be taken to select the most suitable type of net. Dragging large amounts of aquatic vegetation can put great strain on various parts of the net, so it needs to be robust. Circular profile frames are not the most suitable for skimming along the bottom and it is preferable to use one with a straight edge. The author has designed a net frame from aluminium strips readily available from DIY stores. The five sided profile enables the net to be used at various angles ensuring that a leading edge is skimmed through the substrate be it aquatic plants or bottom substrate. The strong mesh material can be purchased from equestrian suppliers. This design has proved especially successful with a number of bottom living species as well as when dragging through actively growing aquatic plants. Alternatively, industry-standard sampling nets used by the Environment Agency can be bought from commercial suppliers.

Larvae should be handled with care due to their small and delicate nature. They can be examined in the palm of the hand either by eye or with a magnifier. A plastic tea spoon can be useful to transfer them. Once out of water the caudal lamellae tend to lie together making examination difficult. In a small rectangular plastic pot filled with water the caudal lamellae splay out and enable one to examine them for characteristic markings and setae.

Rearing larvae in captivity

Larvae can be reared in captivity if a suitable container or small aquarium is prepared with substrate and a food supply. **Care must**

An enhancement of the home made aquarium design is to have a shorter glass front panel than the rear one. This ensures that the rear vertical corners do not appear in photographs as a result of refraction effects through water. Place on a north facing window to prevent overheating. A 'wigwam' made from chopsticks taped together can be added to provide a suitable support when larvae are ready to emerge. A Common Blue Damselfly can be seen emerging (arrowed).

Small Red Damselfly feeding sequence-
Damselfly larvae can be fed in captivity with bloodworms and observed as they eat. The mask is used to hold the prey as it struggles and then guides it to the mandibles where it is masticated. The red colouration of the bloodworm can be seen passing down the gut of the damselfly. Finally the damselfly cleans its mouthparts using the setae on the labium and the mandibles. Note the large eyes used for hunting, which are characteristic of this species.

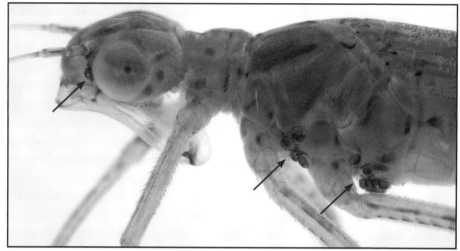

be taken not to use tapwater straight from the tap. Newly hatched larvae will feed on protozoans and then progress to larger food such as water fleas (*Daphnia*), which can be bought from most aquatic suppliers. Bloodworms and mosquito larvae from a water butt provide a good food source for larger larvae. Larger damselfly species can also be fed on very small earthworms or small Frog tadpoles (not Toadpoles which are poisonous).

Glass or plastic aquaria can be bought easily from pet shops and aquarists. Suitable aquaria can be constructed using small plates of thin glass sealed together with silicon mastic. Such tanks can be filled with suitable plants and substrates and enable larvae to be studied and photographed through the front panel.

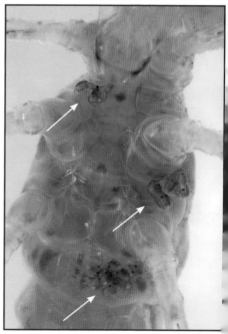

CAUTION - Rearing larvae
Great care should be taken when rearing larvae over winter. A good food supply and warm indoor conditions can lead to premature development and emergence of the adults, when conditions outside are too cold for them to survive.

Larva of Small Red-eyed Damselfly with Mites. Adult damselflies are often parasitised by mites, especially when numbers are very high. Mites are also found on larvae, although they may not be parasitic. The photographs show a larva with mites clustered around the underside of the thorax (arrowed). One mite can be seen just in front of the eye.

Where to find exuviae

When the final instar larva is ready to transform into the adult damselfly, it leaves the water and searches out a suitable support on which to emerge. Several days before emergence, larvae will sit partially out of the water as their breathing mechanism undergoes the change from using oxygen dissolved in the water to air. This can be a good indication that the emergence of larvae is imminent in small aquaria.

Emergence

Damselflies can be loosely referred to as spring species or summer species. In spring species, a somewhat synchronised emergence takes place over a short period of a few weeks. The best example of a spring species is the Large Red Damselfly, which starts to emerge during April in most years and peaks during May. Summer species have a much more prolonged period of emergence over several months when the year's cohort emerges. A good example of this is the Small Red Damselfly, which may still be emerging in late summer (see emergence period histograms on page 21).

Emergence takes place during the day at the water's edge in many species of damselfly, enabling easy observation. Larvae can sometimes be seen swimming towards the bank using their caudal lamellae to propel them forwards. Some species such as Red-eyed Damselfly are strong swimmers. They will then climb vegetation on which to emerge. Some species, such as Common blues, emerge in such large numbers that it is sometime possible to find exuviae piled on top of each where there is a shortage of suitable supports on which to emerge. They may also crawl many metres away from the point at which they left the water before emerging in cover (see below). Others, such as Small Red-eyed Damselfly, will emerge on aquatic plants that break the surface in the middle of a pond or lake. This can make the finding of exuviae quite difficult without the use of waders or a boat.

Emergence of Common Blue Damselfly
Larvae leave the water and may crawl for several metres before emerging.

▲ **New Forest bog seepage**
A good place to look for damselfly exuviae.
Exuviae of Azure, Scarce Blue-tailed, Small Red
and Southern Damselflies have been found on
the emergent vegetation at this site.

When to find exuviae

An exuvia is the ultimate proof of breeding
at a site; warm, sunny, settled periods are
best for emergence and a good time to
look. Most emergence takes place in the
mornings and exuviae are often easier
to find when an emergent damselfly
is present. Being able to find pre-flight
emergents or first flight tenerals is a
good way of finding and confirming
the identification of an exuvia. However,
emergents can be misidentified at this
early stage due to their lack of colour. To

◀ **Mass emergence of Common Blue and
Blue-tailed Damselflies**
Larvae leave the water and can take up every
possible support for emerging.

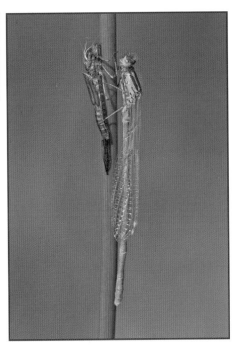

ensure correct identification one should
wait several hours until the colour starts
to develop. Periods of heavy rain / hail
can dislodge them from their emergence
support.

The emergence period offers the best
opportunity to find exuviae and if regular
visits are made to search a site, an accurate
picture of the numbers emerging can
be built up. Frequently counting exuviae
is one of the best ways of assessing the
population size of damselflies. Particular
care should be exercised as exuviae and
any nearby emergent damselflies are small,
fragile and can easily be knocked from their
support or blown or washed away.

**Exuviae are an exact replica of the
final instar larvae.** Take care when
collecting them because some features
may be damaged or distorted during the
emergence of the adult damselfly. It is
sometimes better to remove them along
with their support. They should be placed
into a suitably large container (e.g. a film
canister) and each one labelled with the
location and date of collection. Do not
put more than one in each container, as
damage may occur.

It should be noted that colour is not a
reliable diagnostic feature of damselfly
larvae or exuviae. It can be highly variable
and larvae will often take on the colour of
the substrate in which they are living and /
or the food eaten.

Cleaning of exuviae

Compared to dragonfly exuviae, those
of most damselflies are relatively clean.
However, it may be necessary to clean an
exuvia to enable key features such as the
caudal lamellae to be seen clearly before
attempting identification. Clean them by
soaking them in a small bath of vinegar,
then washing them in water with a drop
of washing up liquid. The caudal lamellae
need to be carefully teased out in water
using a mounted needle and fine forceps.
If more than one specimen is available it
is better to sever them from the abdomen
and prepare them separately. They can
then be dried and examined. Several
soakings may be necessary for this.

Sexing exuviae

Most final larvae and exuviae can be sexed
readily by looking for the presence of

genitalia on the underside of S9 (see earlier section on anatomy). In females this takes the form of a bulbous structure that represents the precursor of the ovipositor. In males two small pointed structures should be visible, although the lack of evidence of an ovipositor is a clear indication.

Reference collection

A reference collection of exuviae provides a useful means of comparing unidentified exuviae with those of known origin. Exuviae should be thoroughly dry and can be stored in film canisters, clear plastic pots or pinned onto cork or pith boards in collection cabinets. Care should be taken when using pins so as not to obscure key features. Caudal lamellae can be separated and mounted on glass microscope slides. Specimens should be clearly labelled with the collector's name, the locality and grid reference and with the date when the exuvia was found.

The use of microscopes

After collection, exuviae can be examined at leisure. The use of a stereo microscope with binocular head will assist in viewing the smaller characteristics that are useful in identification. A zoom magnification is especially useful to quickly change from a macro view then to fine detail.

A trinocular head will also allow a camera to be fitted. Some digital cameras can be connected directly to the computer for quick viewing on a large screen.

If possible, fibre optic illumination is recommended, as it enables the direction of illumination to be adjusted quickly and easily. This can make a big difference when viewing small characters such as setae, where some degree of backlighting highlights such small structures.

Habitat - can provide supporting information and a good clue towards identification. Where appropriate, key habitat details are given under the **Larval Habits** section for each species.

CAUTION: Final Instar Larvae

Some species have distinctive features that enable identification to species in all larval stages, whilst others may only be identified to genus or family.

Only larvae in the final stage can be reliably identified to species level. Many of the features that characterise each species are not fully developed in earlier instars.

Larvae in the final stage can be recognised by the length of the wing buds which cover the fourth abdominal segment.

Take great care when attempting to identify any larvae in which the wing buds are shorter than this unless specifically advised in the species accounts.

All damselfly larvae have ten abdominal segments but the first is often rather difficult to see, so it is best to count from the last segment (S10) backwards.

Photographs used in species sections
Little or no cleaning has been carried out on the larvae and exuviae photographed for the species accounts. It is intended that they should portray the way in which they are found in the field. Obvious debris has been removed with a fine camel hair brush or fine tweezers. In some cases legs may be missing; this is normal and does not prevent an identification. For each species a scale bar indicates actual life size length of body plus caudal lamellae unless stated otherwise.

Emergence periods

The following emergence period histograms show all records by week number for pre-flight emergent adults (Em) and exuviae (Ex) in the Dragonfly Recording Network database up to the end of 2006. These indicate the best periods for finding exuviae.

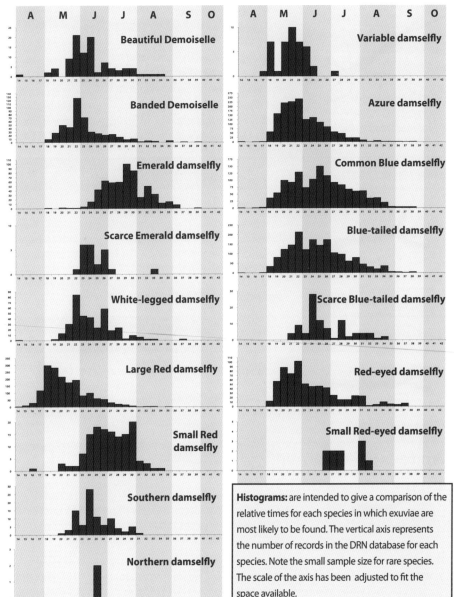

Histograms: are intended to give a comparison of the relative times for each species in which exuviae are most likely to be found. The vertical axis represents the number of records in the DRN database for each species. Note the small sample size for rare species. The scale of the axis has been adjusted to fit the space available.

A quick guide to size and shape

Demoiselles pages 24-29

Banded Demoiselle
Beautiful Demoiselle

Demoiselle larvae and exuviae are large and very distinctive and can be readily identified to genus. They have a **stick-like appearance** with **long spider-like legs** and narrow **rigid caudal lamellae**. The **antennae are long and horn-like** with the first segment as long as the combined length of the other antennal segments. Both British species breed in running water. The combination of the above characters will readily enable identification to genus. *Early instar larvae are distinctive enough to be identified to genus.*

Emerald Damselflies pages 30-35

Emerald Damselfly
Scarce Emerald Damselfly

Emerald larvae and exuviae are large and distinctive in shape and can be readily identified to genus. **Caudal lamellae are rigid and have prominent dark transverse bands. The secondary trachae of the caudal lamellae are perpendicular.** In most but not all species the **labial mask is long and tennis racket shaped.** *Early instar larvae are distinctive enough to be identified to genus.*

Red-eyed Damselfly pages 68

Red-eyed Damselfly

Red-eyed Damselfly larvae and exuviae are large and distinctive in shape and can be readily identified to species. **Caudal lamellae are flaccid and have three prominent dark bands.** N.B. The Small Red-eyed Damselfly is much smaller.

White-legged Damselfly pages 36-37

White-legged Damselfly

White-legged Damselfly has a distinctive profile. The **caudal lamellae have heavily pigmented blotches and terminate with a characteristic long thin filament at the tip.** Occasionally this may be absent.

PLEASE NOTE: The illustrations on these pages are reproduced at life size to allow direct size comparison with larvae and exuviae. NB. They are not in same order as the species accounts.

Distinctive small damselflies

pages 38-45

Large Red Damselfly

Large Red Damselfly larvae and exuviae are distinctive with a squat shape and can be readily identified to species. **Caudal lamellae have a dark patch often forming a X.**

Small Red Damselfly

Small Red Damselfly larvae and exuviae are distinctive with a squat shape and large head. **Caudal lamellae are broad with dark veins and spots.** Only occurs in bogs and associated streams and some fen habitat.

Southern Damselfly

Southern Damselfly larvae and exuviae are distinctive due to small size. **Caudal lamellae are short relative to length of body.** Rare and only found in specialised habitat.

Other small damselflies

pages 46

Azure Damselfly
Variable Damselfly

Northern Damselfly

Azure, Variable, Irish and Northern Damselfly have **characteristic spotting on top hind margins of the head**, which distinguishes them from other species. **Caudal lamellae exhibit subtle differences and need to be studied carefully.** Northern Damselfly has distinctive node on caudal lamellae. Azure and Variable Damselfly larvae and exuviae may prove difficult to separate.

Blue-tailed Damselfly
Scarce Blue-tailed Damselfly

Blue-tailed Damselflies have no spots on head. Caudal lamellae taper to terminate in a point. Blue-tailed is very common in most habitat types. Scarce Blue-tailed is rare and restricted to shallow runnels and seepages with sparse vegetation. It lacks banding on legs.

Common Blue Damselfly

Common Blue Damselfly has no spots on hind margin of head. Caudal lamellae typically with 1, 2 or 3 dark perpendicular lines when viewed from side.

Small Red-eyed Damselfly

Similar to other small damselfly larvae. Caudal lamellae with three pigmented blotches. Look for rows of fine setae on underside of S2 and 3.

In the species section where a larva or exuvia is shown, the scale bar indicates actual life size of body length (from front of head to end of abdomen) plus the caudal lamellae (CL). ⊢——⊣

Beautiful Demoiselle   *Calopteryx virgo*

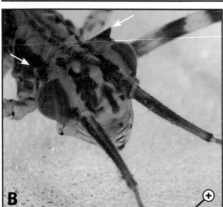

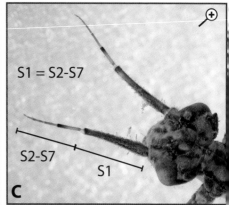

Figure A. Oblique top view of larva. Bar indicates life size length. This larva clearly shows a single pale vertical band at the mid point of the caudal lamellae that is characteristic of this species.

Figure B. Close up view of larval head showing prominent pointed 'tooth' on the top side of the head behind the eyes (arrowed).

Figure C. Top view of head showing characteristic antennae of Demoiselle spp. The 1st segment S1 is as long as all other segments S2-7 combined. N.B. This also applies to Banded Demoiselle.

Figure D. Top view of caudal lamellae on a larva with a single pale vertical band at the mid points. When viewed from this angle the different thickness of the two outer lamellae is apparent.

Size of final instar larvae and exuviae
Length 17-23 mm. CL 6-10 mm.
Demoiselle larvae and exuviae have a
distinctive stick-like appearance with long
spider-like legs (**A**). *Individuals can be
identified as Demoiselle spp. in all larval
stages.*

Head
Head well patterned with **prominent and
pointed occipital tooth** behind each eye
(**B, F**). *This can sometimes be difficult to see
in exuviae due to splitting of the head during
emergence.*
The mask from below has a long cleft in the
middle of the broad front section (**G**).
Long 'horn-like' antennae. The first antenna
segment is as long as the combined length of
all other segments (**C**).

Thorax
Legs are distinctly banded in appearance
(**A**) but can appear heavily stained in some
specimens (**E**).

Abdomen
Some individuals can be dark, especially
those from heathland streams (**E**).

Caudal lamellae
Caudal lamellae are long, narrow and
triangular in cross section (**D**). The middle
lamella is shorter than outer pair. They
usually have a **single pale vertical band
(A,D)**. This can sometimes be difficult to

see in dark heavily stained specimens. The
caudal lamellae can sometimes be lost during
emergence.

Larval habits
Breeds in fast flowing streams, usually with
a stony or gravelly based bottom substrate.
Adult females typically oviposit into
underwater plants in flowing waters, often
with some trees on the banks. Such habitat is
often the best indication of which Demoiselle
species is likely to be present.
Emergence takes place on marginal
vegetation and trees.
NB. Whilst the adults of both Demoiselle
species may be seen together, the larval
habitat is quite distinct. Care should be taken
to establish proof of breeding where both
are recorded.

Possible confusion species
The closely related Banded Demoiselle lacks
the pointed occipital tooth behind the eyes
and usually has two pale bands on caudal
lamellae. Banded Demoiselle favours slower
flowing rivers and streams with a muddy
bottom substrate. There are some sites where
both species can be seen together.
The distinctive shape of both Demoiselles
enables identification to genus at any larval
stage .

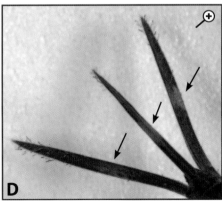

D

Beautiful Demoiselle

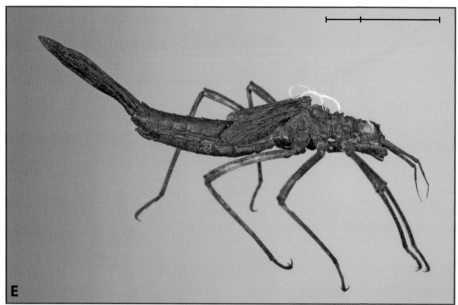

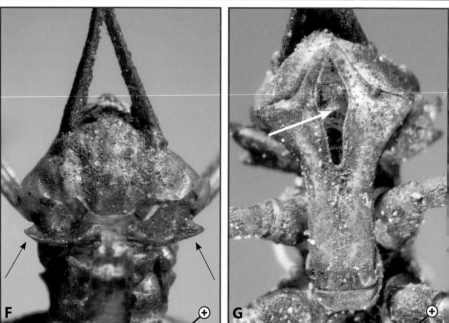

Figure E. Side view of Beautiful Demoiselle exuvia. This specimen is heavily stained. Bar indicates life size length. The characteristic antennae of Demoiselle spp. show well.
Figure F. Close up view of exuvia head showing prominent pointed 'tooth' behind each eye.
Figure G. Underside of head showing characteristic mask with deep central cleft (arrowed).

Banded Demoiselle   *Calopteryx splendens*

A

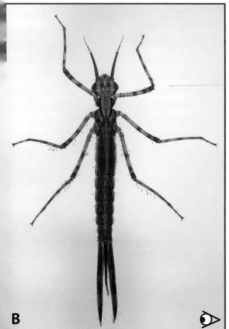

B

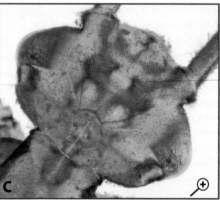

C

Figure A. Oblique top view of larva. Bar indicates life size length. Note the relatively longer caudal lamellae compared to Beautiful Demoiselle.

Figure B. Top view of larva showing 'horn like' antennae.

Figure C. Top view of head showing rounded side behind eyes and lack of pointed tooth.

Size of final instar larvae and exuviae
Length 18-24 mm. CL 9-14 mm.
Demoiselle larvae and exuviae have a
distinctive stick-like appearance with long
spider-like legs (**A, B**). *They can be identified
as Demoiselle spp. in all larval stages (**E**).*

Head
The occipital tooth behind eyes is **less
conspicuous and more rounded (C, D)**
compared to Beautiful Demoiselle.
Long 'horn-like' antennae (D). The **first
antennal segment is longer than the
combined length of all other segments**.
The mask as seen from below has a long cleft
in the middle of the broad front section.

Thorax
Legs are distinctly banded in appearance but
banding can be obscured on specimens from
muddy habitats.

Abdomen
No distinctive features.

Caudal lamellae
Caudal lamellae are long, narrow and
triangular in cross section (**F, G**). The middle
lamella is shorter than outer pair. Usually
with a **two pale vertical bands**. In some
specimens the banding may be difficult to
see. The caudal lamellae are longer relative to

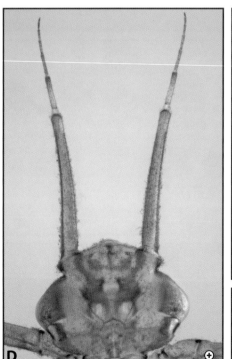

D

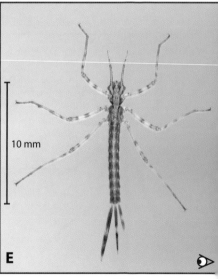

E

Figure D. Top view of head of final stage larva
with rounded hind margin and characteristic
antennae.
Figure E. Top view of early instar larva showing
characteristic body shape.

the body length than in Beautiful Demoiselle (**A, H**).

Larval habits
Breeds in slow flowing rivers and streams, typically where mud and silt accumulate on the bottom. Larvae are best found by searching areas with abundant emergent plants such as root masses that overhang into the watercourse.
NB. Whilst the adults of both Demoiselle species may be seen together, the larval habitat is quite distinct. Care should be taken to establish proof of breeding where both are recorded.

Possible confusion species
The closely related Beautiful Demoiselle has a prominent and pointed occipital tooth behind the eyes and usually has one pale band on caudal lamellae. Beautiful Demoiselle favours faster flowing rivers and streams with a stony or pebble bottom substrate. There are some sites where both species can be seen together.
The distinctive shape of both Demoiselles enables identification to genus at any larval stage (**H**).

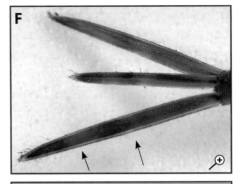

Figure F. Caudal lamellae of larva. The two pale bands are indistinct (arrowed).
Figure G. Caudal lamellae from exuvia. Note the triangular outer lamellae characteristic of this genus.
Figure H. Side view of exuvia showing relatively long caudal lamellae and characteristic body shape.

anded Demoiselle

Emerald Damselfly   *Lestes sponsa*

A

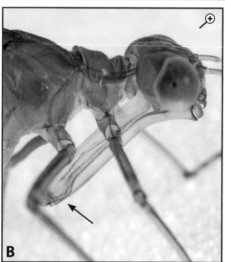

B

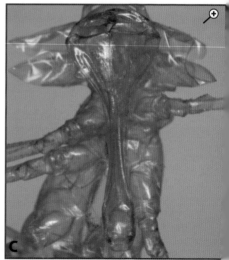

C

Figure A. Side view of larva showing distinctive shape. Scale bar indicates life size. Caudal lamellae show secondary trachae perpendicular to main one. Note the characteristic three dark bands.
Figure B. Side view of head of larva showing distinctively long labial mask, projecting well under the thorax (arrowed) and translucent appearance.
Figure C. Underside of head of exuvia showing characteristic racket shaped labial mask.

30

Emerald Damself

Size of final instar larvae and exuviae
Length 16-22 mm. CL 9-10 mm.
Final stage larvae are larger than most other damselflies and have a characteristic outline, with a slender body, relatively broad head and long lamellae (**A**).

Head
The mask is long and shaped like a tennis racket (**C**). The hinge extends beyond the base of the hind legs i.e. well under the thorax (**B**). The labial palps have **two** setae (requires high magnification).
The head is broad with antennae that are long and thin (**A**).

Abdomen
The **female ovipositor is prominent but does not extend beyond the end of S10 (F)**.

Caudal lamellae
Caudal lamellae have prominent dark transverse bands, usually three (**A, D**). The two outer caudal lamellae are parallel-sided for their entire length and do not taper markedly towards the tip (**D**).
The secondary tracheae are at right angles to the main tracheae (**A, D**).

Larval habits
This species breeds in pools with dense emergent vegetation including in temporary pools that dry out during late summer. Eggs overwinter in diapause and hatch in the spring, followed by rapid larval development before emerging during the summer.
Can only be found reliably as larvae from March.

Possible confusion species
Scarce Emerald Damselfly is very similar and may occur in the same shallow pools. It can be distinguished by more pointed and tapering caudal lamellae. Females of Emerald Damselfly can be identified by the relatively shorter length of the ovipositor.

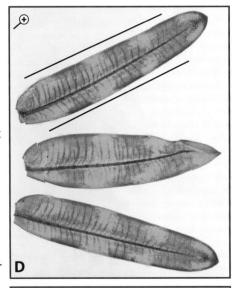

D

Figure D. Caudal lamellae from exuviae showing three characteristic dark bands. Note the near parallel sides of the outer lamellae, which distinguish it from Scarce Emerald.
Figure E. Top view of head of exuvia showing the broad appearance with smoothly rounded hind margin.

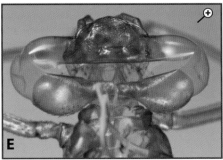

E

Emerald Damselfly

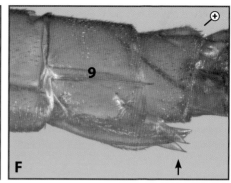

Figure F. Side view of female exuvia of Emerald Damselfly showing last abdominal segments with **ovipositor not extending beyond S10**. In Scarce Emerald ovipositor extends beyond (see page 34).

Figure G. Side view of exuvia of Emerald Damselfly. Scale bar indicates life size. Note the relatively long legs and characteristic parallel-sided caudal lamellae. Larvae of Red-eyed Damselfly can appear superficially similar.

Willow Emerald Damselfly   *Lestes viridis*

Very rare occasional migrant. One exuvia has been found from the north Kent marshes. Final stage larvae are similar to the Emerald Damselfly with a characteristic 'Emerald' outline, with a slender body, relatively broad head and long lamellae.

Shape of mask is distinctive being shorter than other Emeralds, gradually widening and triangular in shape. This separates the species from all other Emerald damselflies. *Any Emerald with a shorter mask that is not racket-shaped should be critically examined and referred for expert opinion.*

This species is associated with overhanging branches, where eggs are laid, giving rise to scarring on the branches.
Length 16-19 mm. CL 7-9 mm.

Figure A. Side view of exuvia. Superficially very similar to Emerald Damselfly with parallel sided caudal lamellae.

Figure B Underside view of shorter triangular shaped mask of exuviae.

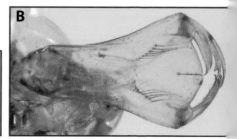

Scarce Emerald Damselfly   *Lestes dryas*

A

B

C

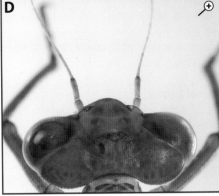

D

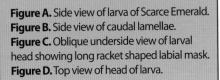

Figure A. Side view of larva of Scarce Emerald.
Figure B. Side view of caudal lamellae.
Figure C. Oblique underside view of larval head showing long racket shaped labial mask.
Figure D. Top view of head of larva.

Size of final instar larvae and exuviae
Final stage larvae of emerald damselflies are larger than most other damselflies and have a characteristic outline (**A**). The body of the female of this species is more robust than other emerald damselflies.
Length 17-22 mm. CL 9-10 mm.

Head
The head is broad (**D**) with antennae that are very long and thin. The labium is long and shaped like a tennis racket (**F**). The hinge extends beyond the base of the hind legs (**C**).
The labial palps have **three** setae (requires high magnification).

Abdomen
The **female ovipositor extends beyond the end of S10 (E)**.

Caudal lamellae
Caudal lamellae have prominent dark transverse bands. The middle lamella usually has two dark bands. The lateral caudal lamellae taper from the middle to the tip.
The secondary tracheae are at right angles to the main tracheae (**B, G**).

Larval habits
This species breeds in pools and ditches with dense emergent vegetation. It is often present at seasonal pools that dry out during late summer. As the name suggests, this species is scarce. It is restricted to Eastern England where it occurs in shallow Pingo ponds in Breckland and coastal marshes in Essex and Kent. It also occurs in Ireland.
It is well adapted to survive in seasonal pools and ditches. Eggs overwinter in diapause and hatch in early spring, followed by rapid larval development before emerging during the summer. As a consequence larvae can only be found reliably during the first half of the year.

Possible confusion species
Emerald Damselfly is very similar and best distinguished from Scarce Emerald by the shape of the lateral caudal lamellae, which are parallel-sided for their entire length. Specimens that have lost their caudal lamellae may prove difficult to separate. The relative length of the ovipositor in females is also distinctive. Females of Scarce Emerald can be identified by the relatively longer length of the ovipositor which extends beyond S10 (**E**).
Final stage larvae of Red-eyed Damselfly are similar in size and also have distinctly banded caudal lamellae. Close examination of the head and caudal lamellae will distinguish the two genera.

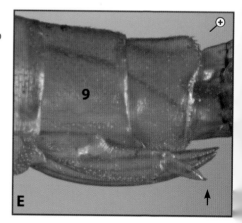

Scarce Emerald Damselfly

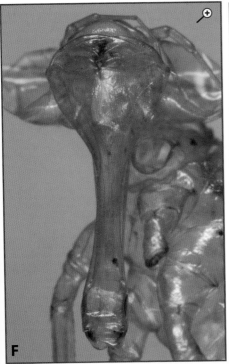

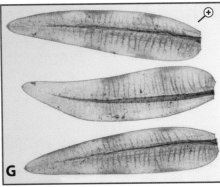

Figure E. Side view of female exuvia showing last abdominal segments with ovipositor extending well beyond S10 (arrowed). In Emerald Damselfly the ovipositor is shorter.
Figure F. Underside view of exuvia head showing long racket shaped labial mask.
Figure G. Caudal lamellae from exuvia showing strong tapering towards the tip.
Figure H. Side view of exuvia, clearly showing strongly tapering lamellae.

Southern Emerald Damselfly   *Lestes barbarus*

Very rare occasional migrant.
Final stage larvae are similar to the Scarce Emerald Damselfly with relatively broad head and long lamellae.
Labium is long and 'racket shaped'. The hinge extends beyond the base of the hind legs. Caudal lamellae very similar in shape to Scarce Emerald Damselfly with which it could be confused. *Due to its rarity, if larvae or exuviae are found at a site where adults are known to occur they should be retained and submitted for expert opinion.*
Length 17-18 mm. CL 7-8 mm.

White-legged Damselfly *Platycnemis pennipes*

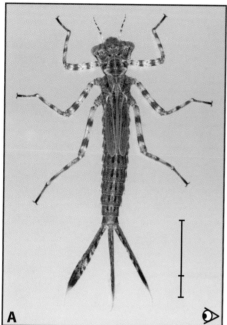

Figure A. Top view of larva.
Figure B. Top view of head. See text for details.
Figure C. Oblique side view of larva showing distinctive outline. Scale bar indicates life size.
Figure D. Caudal lamellae showing long hairs.
Figure E. Caudal lamellae with darker markings.
Figure F. Side view of exuvia.
Figure G. Mask of exuvia showing straight row of 4 setae on prementum (arrowed).
Figure H. Top view of exuvia head showing angular hind margin (arrowed in white).
Figure I. Top view of abdomen of exuvia showing prominent lateral spines.

Size of final instar larvae and exuviae
Length 13-15 mm. CL 6-7 mm, excluding filament. Body has a stout appearance (**A**).

Head
In top view the **hind margin of the head (the occiput) has a distinct angle (B, H,** arrowed in white) and often with pale spots. The head appears to have a 'neck' (**B**). Eyes of larvae with dark vertical lines when viewed from front. **Prementum has a straight row of 4 setae** on the inner surface (**G**). This is a useful feature for confirmation especially when caudal lamellae are missing.

Thorax
The prothorax (see page 2) has distinct bulges (**B, H,** arrowed in black) and is often covered in pale spots in larvae. Legs are distinctly banded. Femur with two prominent dark bands plus two fainter ones at each end(**A**).

Abdomen
The lateral margin has a distinctive outline with prominent spines on S5-9 (**I**).

Caudal lamellae
Caudal lamellae are distinctive with varying pigmented blotches either side of thick central trachea (**D, E**). They terminate with a **characteristic long thread-like filament at the apex** (may sometimes be missing). The margins are covered in long hairs that often collect debris.

Larval habits
Larvae can be found along well vegetated margins of slow flowing waters. They favour stretches with lush vegetation and where roots of emergent plants grow out into the water. Occasionally occur in similar micro habitat at still water sites. Following emergence of adults the exuviae can be found clinging to emergent vegetation and tree roots.

Possible confusion species
The body outline, head shape and distinctive caudal lamellae should enable this species to be separated from all other species.

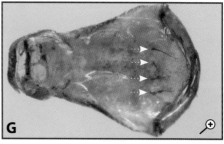

G

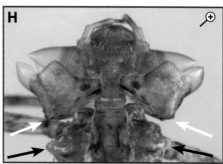

H

S6

I

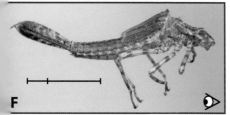

F

Large Red Damselfly   *Pyrrhosoma nymphula*

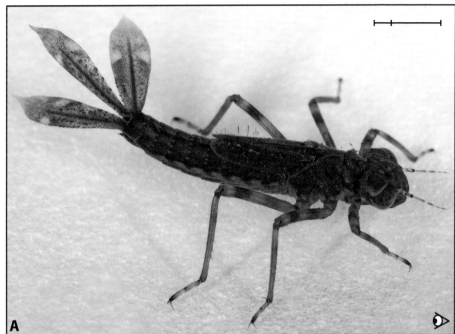

A

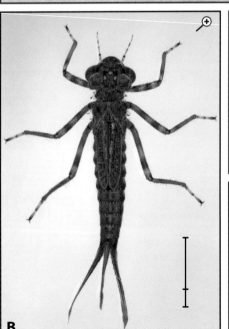

B

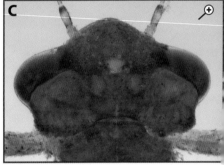

C

D

Size of final instar larvae and exuviae
Length 13-14 mm. CL 4-6 mm.
The larvae and exuviae tend to have a short, squat appearance. From the side, larvae are chunky, with relatively long wing-sheaths and short abdomen (**A**).

Head
The head shape is distinctive. The **back of the head is rectangular** with the hind margin forming a straight line (**C**).

Thorax
In final instar larvae the **wing sheaths project nearly to S6** giving the body a short appearance. On the legs the femora typically have two dark bands (**A, B**).

Abdomen
The abdomen is relatively short , giving this species a squat appearance.

Caudal lamellae
Caudal lamellae are broad, often with **distinctive dark markings and spotting**. These can be highly variable but are typically dark and have an X shaped marking (**E**). Occasionally the markings can be formed by a block of colour.
Setae are present along both margins., those on the apical half being less pronounced.
Larvae living in iron rich water can often collect iron deposits on the lamellae that then dry on exuviae (**I**).

Larval habits
This widespread species breeds in a range of still and running waters. Larvae often become encrusted with debris and can also become heavily stained.
They emerge low down on emergent and bankside vegetation.

Possible confusion species
Larvae and exuviae of Large Red Damselfly are quite distinctive if the dark markings on caudal lamellae can be seen. When lacking, however, this species can be confused with the smaller Small Red Damselfly.

E

F

Figure A. Side view of larva. Scale bar indicates actual length of larva.
Figure B. Top view of larva.
Figure C. Top view of head showing rectangular hind margin.
Figure D. Larva showing caudal lamellae with a light marking.
Figure E. Larva showing caudal lamellae with a dark X-like marking.
Figure F. Side view of larva showing colour variation due to staining.

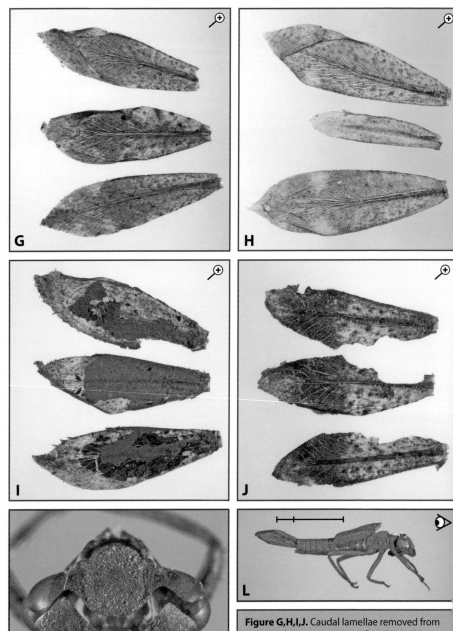

Figure G,H,I,J. Caudal lamellae removed from exuviae showing variation of markings. I is covered in iron deposit.
Figure K. Top view of head of exuvia showing rectangular shape with angular hind margin.
Figure L. Side view of exuvia. Note the relatively short abdomen.

Large Red Damself

Small Red Damselfly   *Ceriagrion tenellum*

A

B

C

D

Figure A. Side view of larva of Small Red Damselfly. Scale bar indicates actual length of larva and caudal lamellae.
Figure B. Top view of larva.
Figure C. Side view of early instar larva.
Figure D. Top view of head showing relatively large eyes and angular rear margin (arrowed).

Size of final instar larvae and exuviae
Length 11-12mm. CL 4-5mm.
The larvae are small with short caudal lamellae. *Early instar larvae show similar characteristics to the final instars.*

Head
A distinctive head shape with the hind margin (**D**) angular, but not as distinctly rectangular as Large Red Damselfly. The head also has prominent eyes and is relatively large compared to the size of the body.
The **inner surface of the prementum has one seta, rarely two (or none) on each side.** With optimal lighting these can been viewed

through the mask on living larvae without harming them (**E**).

Abdomen
Relatively short and squat in appearance.

Caudal lamellae
The caudal lamellae are broad and relatively short compared to the length of the abdomen (**A, B, C**). They have a distinctive shape marked with irregular blotches and are slightly pointed (**F, G**). The primary tracheae are thick at the base giving rise to prominent secondary branches most of which slant strongly backwards. The lamellae are divided by a indistinct diagonal nodal line. One margin of the basal half of each lamella has stout setae reaching the mid point (**F**). The setae on the other margin reach to about one third. The distal part has delicate setae that are only visible under a microscope.

Larval habits
This species breeds in bogs and fens and slow flowing streams associated with such habitat.
Larvae can be found in areas of submerged bog moss *Sphagnum* spp. and other bog plants.
In captivity larvae have been observed to

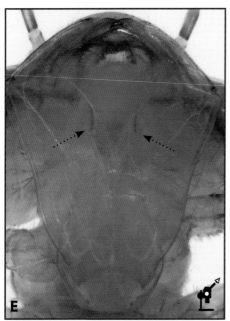

Figure E. Underside of mask on larva showing characteristic setae (arrowed).
Figure F. Caudal lamellae on larva showing irregular blotching and stout setae on basal margins only.

Small Red Damself

intermittently flick their lamellae, possibly as a display.

They are often found in association with larvae of Large Red Damselfly and in a few suitable areas with Southern Damselfly.

Possible confusion species

Larvae and exuviae of Large Red Damselfly are a similar shape but slightly larger. The caudal lamellae can appear similar to those of lightly marked Large Red Damselflies though the distinctive dark veins in clean Small Red Damselfly are more prominent. Where it occurs with Southern Damselfly, the larvae and exuviae can be similar in size but the broader shape of the caudal lamellae of Small Red Damselfly quickly distinguish the two.

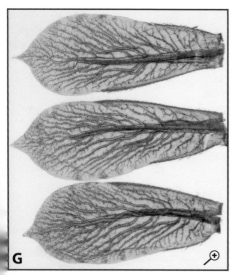

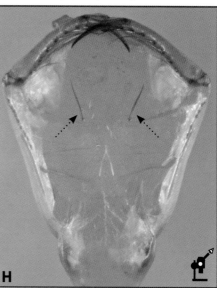

Figure G. Caudal lamellae separated from exuvia showing dark backward facing veins.
Figure H. Underside view of mask from exuvia showing characteristic two setae (arrowed).
Figure I. Side view of female exuvia.

Southern Damselfly   *Coenagrion mercuriale*

A

B

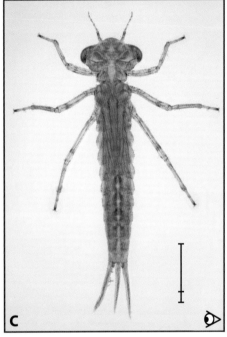

C

E

F

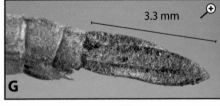

3.3 mm

G

D

H

Size of final instar larvae and exuviae
The **larvae and exuviae are smaller** than most other species.
Length 11-14 mm. CL 3-4 mm.

Head
The hind margin of the head is curved. The upper surface of the back of the head lacks the characteristic spotting of other *Coenagrion* species (**D**).
The antennae have 7 segments.

Caudal lamellae
The **caudal lamellae are short** relative to the body length (**A, B, C**) and have no distinct markings (**E, G, I**). They taper to a point with no obvious node or stout setae. After the mid point the setae take the form of denser, longer hairs. This is especially noticeable in larvae, which often collect fine debris (**E**).

Larval habits
This scarce species breeds in relatively open and unshaded stretches of flowing waters in two habitat types in Britain. It occurs in shallow base rich, heathland streams, flushes and valley mires where typical plants include Marsh St John's-wort *Hypericum elodes* and Bog Pondweed *Potamogeton*

elodes.. Large populations are also associated with calcareous water meadow ditch systems along the Itchen and Test valleys in Hampshire, where typical plants include Watercress *Rorippa nasturtium-aquaticum*, Fool's Watercress *Apium nodiflorum* and Brooklime *Veronica beccabunga*. The species needs warm sites, or at least those where water temperature is relatively high in winter, such as those associated with springs or other groundwater sources.

The Southern Damselfly exhibits a two-year life history in Britain. Adults emerge from their final larval stage between mid-May and late July. They usually emerge in the morning. Final instar larvae leave the water by ascending emergent vegetation, rather than by walking onto shore. There appears to be no consistent trend in plant species used but plants with a rigid stem such as rushes *Juncus* sp., are frequently used.

Possible confusion species
Larvae and exuviae are small with short caudal lamellae. However, Small Red Damselfly also has relatively short caudal lamellae and can be found in the same habitat. That species has a relatively larger head with a distinctly angular hind margin and the caudal lamellae are distinctly broader in shape.

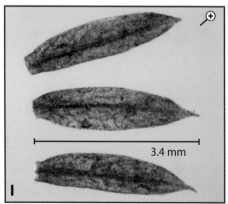

3.4 mm

I

Figures A & B. Side views of larvae showing variation in colour. Scale bar shows actual size.
Figure C. Top view of larva showing the relatively short caudal lamellae.
Figure D. Top view of head, lacks spots.
Figure E. Side view of caudal lamellae of larva showing fringing hairs covered in fine debris.
Figure F. Side view of exuvia.
Figure G. Side view of caudal lamellae of female exuvia.
Figure H. View of mask from below.
Figure I. Side view of caudal lamellae separated from exuvia.

Northern Damselfly   *Coenagrion hastulatum*

A

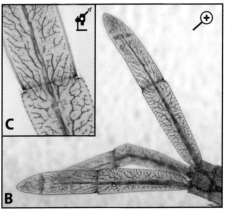

B

C

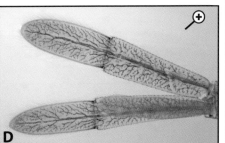

D

E

6
4
2

Size of final instar larvae and exuviae
Length 14-15 mm. CL 5-6 mm.
Body colour is variable, mainly green and brown. **Found in Scotland only**.

Head
Antennae with 6 segments (E). Spotting behind the head is regular and widely distributed (**E,F**). Labial palps with 4 setae.

Thorax
Legs are distinctly banded on femora (**A**).

Abdomen
Abdomen is spotted but not distintinctively.

Caudal lamellae
Caudal lamellae have a very distinct and darkened nodal constriction and a dark perpendicular nodal line (B,C,D). The

furthest half from the body is fringed with dense fine hairs. Some individuals also have 1 or 2 narrow dark lines towards the rounded tip (**B**).

Larval habits
Breeds in acidic, shallow, sheltered pools and margins of lochs in central Scotland. Most sites are fish free and larvae are often found in open water clinging to underwater plants. Adult females typically oviposit into floating and emergent plants. The larvae take two years to develop. Emergence takes place on marginal vegetation, especially sedges and Water Horsetail.

Possible confusion species
Northern Damselfly is one of a number of species with similar larvae. The prominent spotting on the head distinguishes it from Common Blue and Blue-tailed Damselflies, with which it commonly occurs. The Azure and Variable Damselflies are also very similar, but lack the dark perpendicular nodal line on the lamellae.

Figures A. Side view of larva. Scale bar shows actual size.
Figure B. Caudal lamellae of larva.
Figure C. Close up view of node of caudal lamella showing dark nodal constriction.
Figure D. Variation on caudal lamellae.
Figure E. Top view of head showing dark spotting.
Figure F. Top view of head in less intensely spotted individual.
Figure G. Side view of exuvia. Scale bar shows actual size.
Figure H. Caudal lamellae from exuvia.

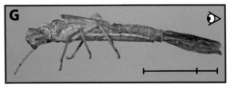

G

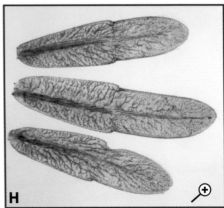

F

H

Azure Damselfly   *Coenagrion puella*

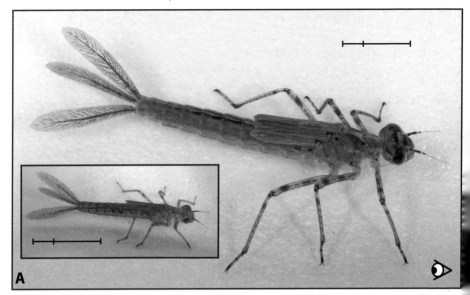

Size of final instar larvae and exuviae
Length 12-13 mm. C L 5-6 mm.
Body colour is variable yet mainly green and brown (**A**).

Head
Antennae with 7 segments (**B**). **Head with prominent distinctive, spotting behind the eyes (B)** similar to Variable Damselfly. The mask has **two rows of setae inclined to each other at an angle (Ø) of less than or equal to 90⁰** (**H**), *(In Variable Damselfly the angle is greater than or equal to 90⁰).*

Thorax
Legs are distinctly banded in appearance with two dark bands on each femur (**A**).

Abdomen
The abdomen is spotted but this does not distinguish it from other species.

Caudal lamellae
Caudal lamellae are nodate terminating in a pointed tip (**C, D, E, I, J**). The nodal line can sometimes be lightly pigmented and can vary from almost perpendicular to oblique. The length of the caudal lamellae is typically >4x its width at the node (measured on fully developed mid lamella). Where they attach to the abdomen the lamellae are narrower with a straighter taper compared to Variable Damselfly (**C**).

Larval habits
Breeds in ponds, well vegetated lakes and rivers and canals. Adult females typically egg lay whilst **in tandem**, into underwater plants growing at the water's surface. Emergence takes place on emergent and marginal vegetation.

Possible confusion species
Azure Damselfly is one of a number of species with similar larvae. The prominent spotting on the head distinguishes it from Common Blue and Blue-tailed Damselflies, with which it frequently occurs. Azure Damselflies with pigmented nodal line can initially be confused with Common Blue which lacks spotting on top of head. The Variable Damselfly is very similar and one should look closely at the fine detail described for these species. Variable Damselflies tend to be brown with a rounded tip to broader caudal lamellae.

E

Figures A. Side view of final instar larva with green colouration. Scale bar shows actual size. Inset shows final instar larva with brown colouration.
Figure B. Top view of head of larva showing characteristic dark spotting.
Figure C. Caudal lamella showing straight taper to abdomen. Note the pointed tip.
Figure D. Variation on caudal lamellae with broader less pointed tip. N.B. This lamella is similar to Variable Damselfly.
Figure E. Variation on caudal lamellae with more pointed tip.

F

Figures F. Side view of Azure Damselfly exuvia (Scale bar shows actual life size).
Figure G. Top view of head showing lighter spotting (compared to Fig. B).
Figure H. Top view of mask removed from exuvia to show setae. NB. The dotted lines at the base of the setae are at less than 90^0 ($\emptyset$).
Figure I. Caudal lamellae removed from exuvia derived from a green larva.
Figure J. Caudal lamellae removed from exuvia derived from a brown larva. The nodal line is lightly pigmented.

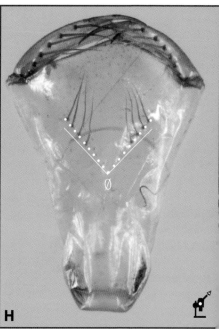

H

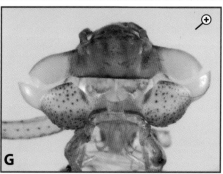

G

I

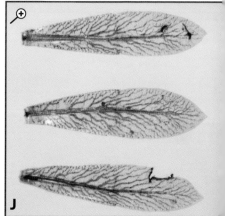

J

Caudal lamellae of Azure and Variable Damselflies - special notes

The Variable Damselfly is highly variable both as an adult and in the larval stages. A number of features, including the caudal lamellae and labial mask (see page 57), need to be carefully examined before the identification of Variable Damselfly larvae and exuviae can be confirmed.

Caudal lamellae should be examined and measured under a microscope. The lamella of larvae can be removed with fine forceps without detrimental effect. A fully developed middle lamella should be used.

(i) Measure the width across the mid-lamella from each margin where the coarse setae end (y).
Measure the length from the mid point of this line where it crosses the main vein to the tip (x).
Calculate the ratio x/y between the two.

(ii) Draw the outline of the lamella with particular attention to the shape of the tip; Is it rounded? Or is it noticeably or slightly pointed?.

(iii) Compare the ratio and tip profile against the table opposite which shows the outlines of middle lamella of both species.

Azure	Ratio x/y	Variable
	1.83	
	1.94	
	1.94	
▼ The profile of caudal lamellae of Azure Damselfly shown below tend to be narrower with a markedly pointed tip.	2.00	
	2.03	
	2.04	
	2.07	
	2.13	
	2.27	▲ The profile of caudal lamellae of Variable Damselfly shown above tend to be broader with a rounded tip.
	2.29	
	2.43	N.B. The x/y ratio overlaps between the two species.
	2.56	The tip shape may separate them,
	2.70	although a few will prove inconclusive. Now go to page 57.

(i)

(ii)

Variable Damselfly   *Coenagrion pulchellum*

A

B

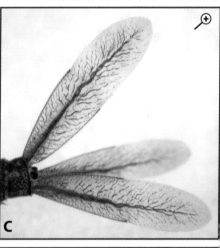

C

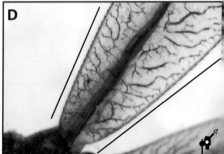

D

Size of final instar larvae and exuviae
Length 13-14 mm. CL 5-6 mm.
As the name suggests this species is
highly variable.
*A number of features need to be carefully
assessed before confirming this species'
identification (see pages 51 and 57).*
Whilst larvae are predominantly dark
brown (**A**), the colour can vary through
light brown to green (**B**).

Head
**Head with prominent, distinctive
spotting behind the eyes (B)**. Antennae
with 7 segments *(both features are similar
to Azure Damselfly)*.
**The mask has two rows of setae
inclined at an angle (Ø) to each other
of more than or equal to 90⁰ (J, K L)** *(In
Azure Damselfly the angle is less than or
equal to 90⁰)*.

Thorax
Legs are distinctly banded in appearance

with two dark bands on each femur (**A**).
In final instar larvae the characteristic
lobed shape of the adult prothorax can be
seen through the larval skin (**B**) on some
specimens.

Abdomen
The abdomen is spotted but does not
distinguish it from other species. Larvae
are predominantly brown in colouration.

Caudal lamellae
Caudal lamellae are **nodate with a
rounded or slightly pointed tip (C, E,
F, G, M, N, O, P, Q)**. In some specimens
the tip shape can be similar to the less

Variable Damselfly

pointed lamellae of Azure Damselfly. The nodal line can vary from almost perpendicular to oblique and is often pigmented. The length of the caudal lamellae is typically less than 4x its width at the node (measured on fully developed mid lamella). Where they attach to the abdomen the lamellae are broader and gently curved (**D**) compared to Azure Damselfly.

See special notes on page 51 for comparing measurements of caudal lamellae.

Larval habits
Breeds in well vegetated lakes, dykes and canals. Adult females typically oviposit in tandem into underwater plants. Emergence takes place on marginal vegetation.

Often found in association with Azure Damselfly larvae.

Possible confusion species
Variable Damselfly is one of a number of species with similar larvae. The prominent spotting on the head distinguishes it from Common Blue and Blue-tailed Damselflies, with which it commonly occurs.
The Variable Damselfly is highly variable and some characters on some specimens appear to overlap with Azure Damselfly. It is strongly recommended that a number of characters need to be carefully examined wherever Variable Damselfly is suspected.
The Azure Damselfly is common and widespread and very similar. In the majority of cases where larvae or exuviae are found it will be that species. Look closely at fine details described for these two species whenever Variable Damselfly is suspected.

N.B. To separate Variable Damselfly requires a number of features to be checked, including a knowledge of the species distribution. Searching for adults during optimal weather conditions during the flight season will often indicate what larvae and exuviae to expect.

H

I

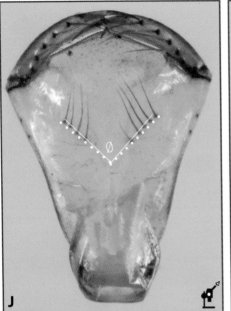

J

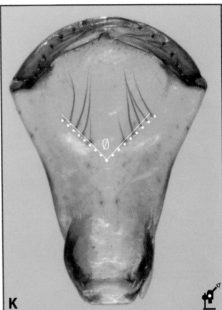

K

Figures A. Side view of final instar larva with dark brown colouration. Scale bar shows actual size.

Figure B. Top view of head and thorax of larva showing characteristic dark spotting on back of head. *N. B. The lobed shape of adult prothorax shows through on this specimen. Compare to Azure Damselfly on page 48.*

Figure C. Caudal lamella on larva showing rounded tip.

Figure D. Detail of middle caudal lamella showing curved taper and broader point of attachment to abdomen.

Figure E, F, G Variations on caudal lamellae with varying tip shape.

Figure H. Top view of head showing dark spotting on a dark exuvia.

Figures I. Side view of exuvia (Scale bar shows actual life size).

Figures J, K, L. Top view of inner surface of mask removed from exuvia to show setae. NB. Dotted lines joining the setae bases are greater than 90° (Ø). Compare with Azure Damselfly on page 50.

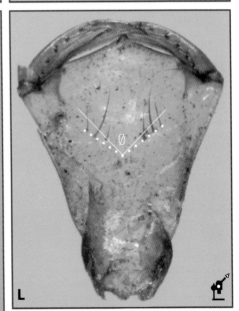

L

N.B. All specimens illustrated here were confirmed by breeding through to emergence to confirm identification.

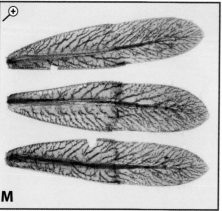

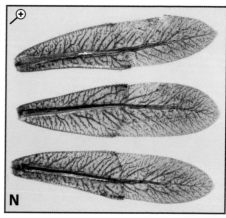

Figures M, N, O, P, Q. Variations of caudal lamellae from Variable Damselfly exuviae taken from sites across the country.
Rounded tips are typically a key characteristic of one or more lamellae.
Lamellae are often broader than those of Azure Damselfly. The *mid lamella is more curved and broader at the base* where it attaches to the abdomen.
Compare these photographs with those of Azure Damselfly on pages 48-50. Please also refer to the key ratios described on page 51.

N.B. Due to the difficulties of identification of Variable Damselfly, all specimens illustrated here were confirmed by breeding through to emergence.

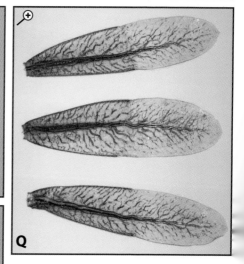

Variable Damselfly

A cautionary note when distinguishing Azure and Variable Damselflies

Various authors have attempted to separate Azure and Variable damselflies using a range of characters. However, there is some overlap between the features of the two species making it difficult to distinguish every specimen reliably.

The caudal lamellae are the most readily available feature (see page 51) and some individuals can be identified by this alone. In combination with **the included angle of the setae on the inner surface of the mask** (see photographs below) the majority of specimens can be identified to species.

Please note that the following features have proven to be of no value to distinguish these two species: Colouration, marginal setae on the caudal lamellae, angle of the nodal line, setae on any abdominal segment, number of twists along the primary tracheae.

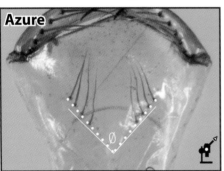
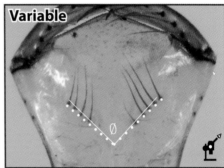

The angle of setae on the inner surface of the mask appears to be a reliable feature. The dotted line on the photographs runs along the base of the setae. The solid lines intersect at 90°. The Azure Damselfly has a lesser angle than 90° (∅), whereas the Variable has a greater angle than 90° (∅).
This feature should be used in combination with the measurements of caudal lamellae on page 51.

Irish Damselfly | *Coenagrion lunulatum*

Exuviae are 14 - 17 mm CL 5 - 8 mm. Spotting behind the head is uneven, sparser on margins.
Antennae with 6 segments.

Caudal lamellae are similar to Northern Damselfly (see pages 46-47).
Currently recorded in Ireland only.

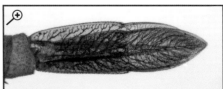

Common Blue Damselfly *Enallagma cyathigerum*

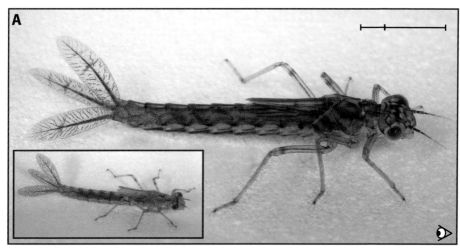

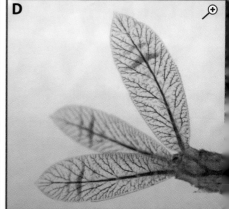

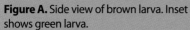

Figure A. Side view of brown larva. Inset shows green larva.
Figure B. Top view of light brown larva.

Size of final instar larvae and exuviae
Exuviae are 14 - 18 mm. CL 6 - 7 mm.

Head
Antennae with 6 segments (occasionally 7).
Lack of spotting on back of head
(**F, G**). This is a useful characteristic to distinguish it from Azure and Variable Damselflies.

Thorax
On the legs there is one dark band on the femur, but not on the tibia (**A, B**).

Abdomen
Overall body colouration is highly variable, ranging from green through to dark brown (**A**). Some individuals are especially heavily marked.

Caudal lamellae
Caudal lamellae are subnodate. Stout setae on both edges reaching between one third and two thirds the length of

the lamella (**C, D, E, I, J, K L, M**). They have **1-3 dark perpendicular stripes**. Occasionally the stripes are absent or vary in intensity. The caudal lamellae are broader, especially at the base, than other blue damselflies and markedly less pointed than Blue-tailed Damselfly. NB. The markings on the caudal lamellae are highly variable. On exuviae the dark perpendicular stripes are sometimes difficult to see as the lamellae often dry together and overlap. These should be removed and examined.

Larval habits
This is a common and wide ranging species. Larvae can be found in a wide range of habitat types, especially large open pools and slow flowing rivers. Specimens from heathland pools can be particularly dark in colouration.
When ready to emerge the larvae leave the water by climbing any available support projecting above the water's surface. During mass emergence exuviae can be found in large numbers piled on top of each other. On occasions they will leave the water and crawl for several metres before emerging.

Possible confusion species
Initially could be confused with any small damselfly, especially Blue-tailed, Azure or Variable Damselfly. The lack of spotting on

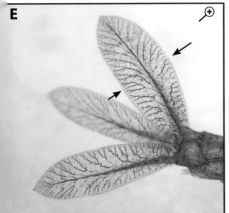

E

> **Variation of caudal lamellae on larvae.**
> **Figure C.** Caudal lamellae showing three distinct lines.
> **Figure D.** Caudal lamellae showing two distinct lines and one lighter.
> **Figure E.** Caudal lamellae showing very faint lines. Such individuals need to be examined carefully. End of stout setae on each margin are arrowed.

ommon Blue Damselfly

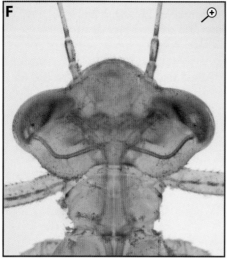

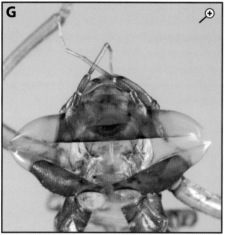

the back of head separates it from Azure and Variable Damselflies.

Larvae without perpendicular bands on caudal lamellae can initially be difficult to separate from Blue-tailed Damselfly. That species often has dark spots at base of wing buds which are always lacking in Common Blue.

Exuviae can be difficult to distinguish from Blue-tailed Damselfly where the caudal lamellae have dried together. The lamellae are obliquely nodate and more pointed in that species. The stout setae end one third to half way (see pages 65 and 67 for comparison).

Small Red-eyed Damselflies often occur in the same habitat. That species has dark blotches that can appear similar to dark lines on caudal lamellae. The rows of stout setae on the underside of abdominal segments S2-3 will distinguish it from Common Blue Damselfly (see pages 71 and 73.)

Figure F. Top view of head of larva showing lack of spots on upper hind margin.
Figure G. Top view of head of exuvia showing lack of spots on upper hind margin.
Figure H. Side view of exuvia. The dark perpendicular lines are just visible on the caudal lamellae.

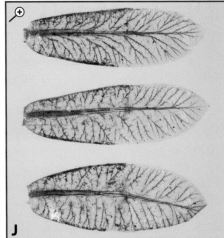

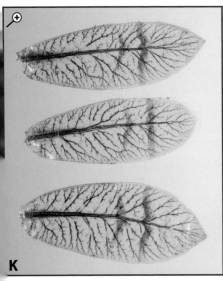

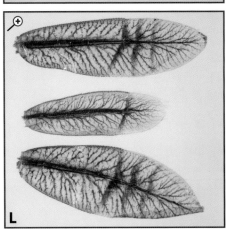

Variation of caudal lamellae from exuviae of Common Blue Damselfly.
Caudal lamellae taken from exuviae collected from one site on the same day. They show variation in dimensions as well as the number of vertical pigmented lines.
Figure I. Pale indistinct lines.
Figure J. Pale 1 line.
Figure K. 2 line.
Figure L. 3 line.
Figure M. 3 line with darker pigmentation.
N.B. The middle lamella has failed to develop to its proper dimensions in **L** and **M**.

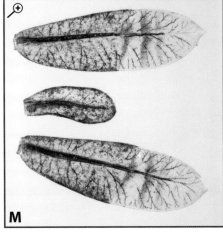

ommon Blue Damselfly

Scarce Blue-tailed Damselfly *Ischnura pumilio*

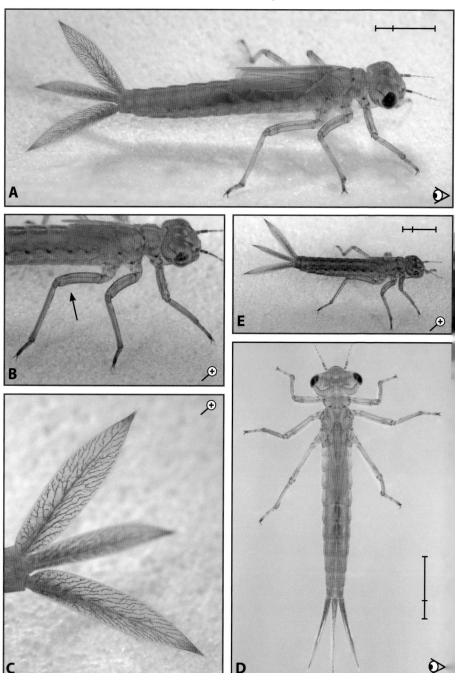

Size of final instar larvae and exuviae
Length 11-12 mm. CL 4-6 mm.
Final instar larvae and exuviae are **smaller** than Blue-tailed Damselfly when seen together for comparison. The overall impression is a one-coloured body lacking any obvious marking, especially on the legs, compared to Blue-tailed Damselfly. This gives the larvae a distinctive appearance.

Head
The back of the head lacks spotting, although the presence of setae can give the impression of spots.

Thorax
The legs have **no or very faint banding** on the femur or tibia. This 'jizz' can be one of the first clues when separating the species (**A,B,D**) and can also be used on early instar larvae (**E**) if not obscured with detritus on body.

Abdomen
The **setae on the lateral ridges are the same thickness and length as other setae** on the underside of the abdomen (**F**). They can be difficult to distinguish especially if the larva or exuvia is covered in debris.

Figure A. Side view of final stage larva. Scale bar indicates life size.
Figure B. Close up of legs showing lack of distinct banding.
Figure C. Caudal lamellae on final stage larva.
Figure D. Top view of larva.
Figure E. Side view of early instar larva. Even at this size the lack of banding on legs and overall mono colouration are visible.
Figure F. Top view of fine setae on lateral ridges that are the same size as other setae.

Caudal lamellae
Caudal lamellae are relatively short tapering from the middle to a distinctive point. They are less transparent than Blue-tailed Damselfly and take on an overall brown ground colouration. Caudal lamellae are dark towards the tip, but tip itself is pale. On exuviae the caudal lamellae should be carefully separated out and cleaned and then viewed with a magnifier to reveal these characteristics.

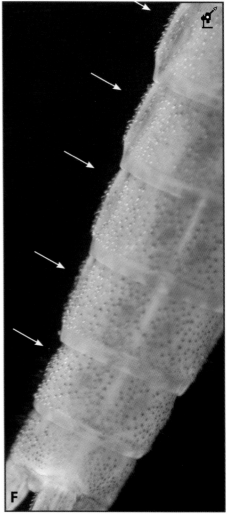

F

G

H

Figure G. Caudal lamellae of early instar larva are more elongate than final instar but still exhibit the uniform colouration.
Figure H. Side view of exuvia. Although covered in debris the legs show lack of banding.
Figure I. Caudal lamellae from exuvia are noticeably shorter than Blue-tailed Damselfly.

Larval habits

This species occurs in restricted micro habitat conditions. **Larvae can be found in very shallow runnels, slow flowing seepages, bog seepages and ditches with little water.** The habitat is typically open with minimal encroachment from vegetation. Larvae are often found in water depth of only several centimetres. It is sometimes found in active earth extraction quarries (chalk, gravel, sand) where shallow runnels occur, such as those created by wheel ruts. It is a transient species that easily disappears with successional habitat changes.

When ready to emerge the larvae leave the water by climbing any available support projecting above the water's surface. Emergence takes place within a few centimetres of the water's surface.

Possible confusion species

Initially larvae and exuviae could be confused with any small damselfly, especially Blue-tailed Damselfly. The smaller size, **lack of banding on legs** and uniform colouration of the caudal lamellae

are good indicators of this species. ***The specialised nature of its shallow water habitat requirements is a strong clue to this species' presence.***

In some areas the Southern Damselfly can also occur at the same site. Size and shape are similar and both have pointed caudal lamellae. In that species the caudal lamellae are noticeably shorter relative to body length.

4.8 mm

I

Blue-tailed Damselfly

Ischnura elegans

A

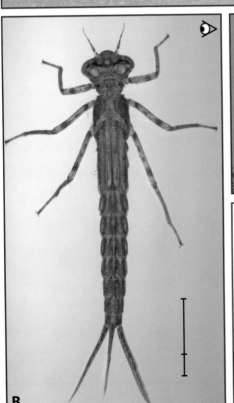

B

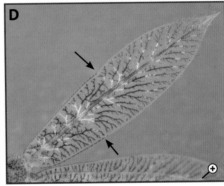

D

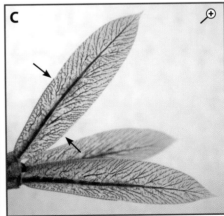

C

65

Size of final instar larvae and exuviae
Length 13-15 mm. CL 5-6 mm.
Highly variable in colour, ranging from dark brown through various shades to bright green.

Head
No spotting on the head. The head appears small relative to the size of the body.

Thorax
Two small dark spots (E) at the wing bases are diagnostic for this species (may be absent in some specimens). They may be obscured by debris and do not persist to the exuviae. These spots are visible in early instar larvae **(F)** before the wing buds start to develop.
On the legs the femora have a dark band.

Abdomen
The **setae on the lateral ridges are larger and thicker than those on the underside**

of the abdomen **(G)** whereas in Scarce Blue-tailed Damselfly they are similar size.

Caudal lamellae
Caudal lamellae are long, thin (length = 4x width) and **distinctly pointed**. They are **obliquely subnodate (C, D, H)** with stout setae reaching the mid point on one side and a third of the way on the other **(D)**. Trachea and branches can be brightly coloured.

Larval habits
Breeds in most freshwater habitats. Adult females typically oviposit alone into any plant material growing in water.
Emergence takes place on marginal vegetation.
In Northern Britain development is semi-voltine i.e. taking 2 years to complete.
In southern Britiain development is uni-voltine i.e. taking one year.
In southern Europe development can be bi- or tri- voltine i.e. with 2 to 3 generations in one year.

Possible confusion species
Common Blue and Azure Damselflies are both superficially similar. Azure damselflies can be distinguished by spotting on top of head. Common Blue Damselflies have much

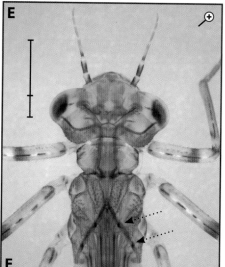

E

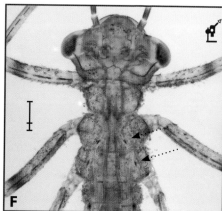

F

broader caudal lamellae often with vertical lines.

The Scarce Blue-tailed Damselfly is similar but lacks banding on the femora. It occurs in runnels and seepages and thus its specialised habitat requirements will preclude it in most cases.

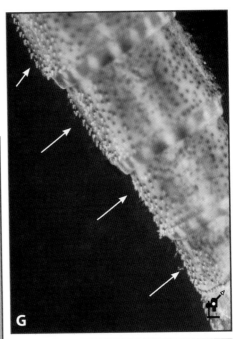

G

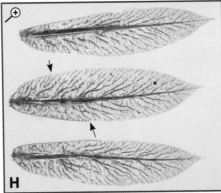

H

Figure A. Side view of final instar larva. Scale bar indicates life size. (Inset shows early instar larva).

Figure B. Top view of larva showing relatively small head.

Figure C. Caudal lamellae on larva showing pointed tip and oblique node (end of coarse setae on each margin arrowed).

Figure D. Caudal lamellae on larva with brightly coloured veins (end of coarse setae on each margin arrowed).

Figure E. Top view of final stage larva showing relatively small head with lack of spotting. Dark spots at wing bases (arrowed) are diagnostic. Note the dark banding on the femora of the legs.

Figure F. Top view of very early instar larva showing dark spots at wing bases (arrowed). If present these are the best way of distinguishing this species from other small damselflies. Wing buds are just starting to grow.

Figure G. Top view of coarse setae on lateral abdominal ridges. NB. that these are coarser than other setae.

Figure H. Caudal lamellae removed from exuvia. Note the tapered, pointed tip.

Figure I. Side view of exuvia.

I

Red-eyed Damselfly

Erythromma najas

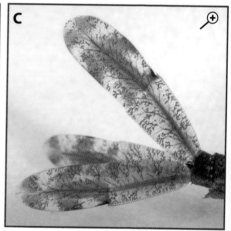

Figure A. Side view of larva. Scale bar indicates life size. The caudal lamellae on this specimen have lighter banding than usual.
Figure B. Top view of larva. Note the relatively small head and long abdomen.
Figure C. Side view of caudal lamellae on larva showing three dark bands between the node and tip. Each lamella has clearly visible and detailed tracheation.

Size of final instar larvae and exuviae
Length 18-22 mm. CL 8-9 mm.
A large species with a distinctively long and banded appearance (**A, B**). Early instar larvae (**E**) can also be identified by this.

Head
The **head is small relative to the length of body** (**B**). The mask is longer and narrower (**F**) than other damselflies, except Demoiselles and Emerald damselflies.

Abdomen
Both Red-eyed damselflies *Erythromma* spp. have **distinctive rows of characteristic setae on the underside of S2 and S3 (G)**.

Caudal lamellae
Caudal lamellae are distinctly **nodate**, which

gives a distinctive two part appearance (**C, H**). The basal part has stout setae whereas the distal part has setae that are barely visible, even with high magnification. The **distal section has three dark transverse bands**.

Larval habits
Larvae can be found in and around emergent and submerged plants. The adults show a strong association with water lilies and other plants with floating leaves. Larvae take two years to develop. Late stage larvae can often be found as early as January.
Emergence typically takes place on aquatic plants rising from the water's surface.

> **Figure D.** Oblique top view of head of larva. The slight spotting is due to the presence of setae.
> **Figure E.** Side view of early instar larva (Note the wing buds are just starting to grow). The caudal lamellae on this specimen make it readily identifiable to this species.

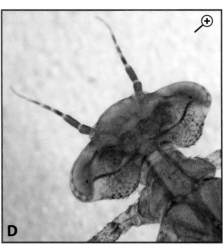

D

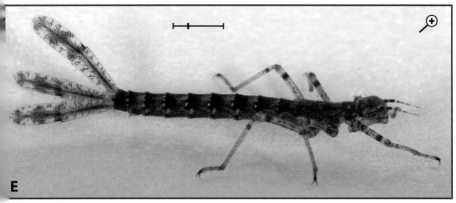

E

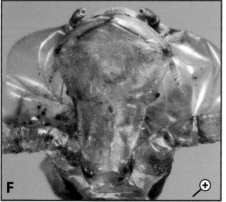

F

Possible confusion species

None, if caudal lamellae can be viewed clearly. The distinct dark banding and rounded lamellae are characteristic. The larger size of final instar larvae should separate them from most other species. Emerald Damselflies can appear similar in size, but the tapering lamellae are quite distinctive with perpendicular secondary veins and the racket shaped labial mask.

Figure F. Underside of head of exuvia showing relatively long, narrow mask.
Figure G. Underside view of exuvia showing characteristic rows of distinctive setae on S2 & S3.
Figure H. Magnified view of caudal lamellae from exuvia showing distinct node and setae on basal half. The three dark bands are characteristic of this species.
Figure I. Side view of exuvia.

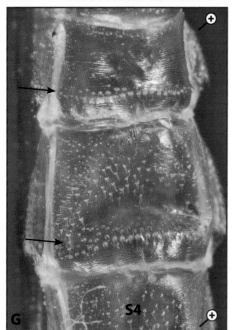

G   S4

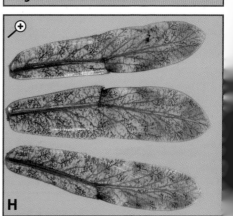

H

I

Red-eyed Damselfl

Small Red-eyed Damselfly *Erythromma viridulum*

A

B

S4

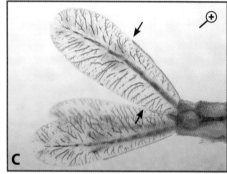

C

D

Figure A. Side view of final stage larva. Scale bar indicates actual life size.
Figure B. Underside of abdomen showing characteristic rows of setae on S2 & S3 (arrowed).
Figure C. Caudal lamellae with characteristic blotches from node towards the tip. The points at which stout setae end are arrowed.
Figure D. Lamellae with darker blotches than usual.

Size of final instar larvae and exuviae
Length 12-14 mm. CL 5-6 mm.
This species is **smaller and less robust than Red-eyed Damselfly**.

Head
The mask is relatively long and narrow (**G**).

Abdomen
The overall body colour is bright green with some individuals taking on shades of brown. **Rows of fine setae on the underside of S2 and S3 (B, H)** are characteristic of Red-eyed damselflies *Erythromma* spp. When examining larvae the rows of setae maybe obscured by a film of water (**F**). Specimens should be temporarily dried with tissue to reveal the setae.

Caudal lamellae
Caudal lamellae are distinctly rounded with three blotches (C, D). In some specimens these may be faint or not visible.
The lamellae are **nodate**, with stout setae to the mid point on one margin and approximately one quarter of the other (**C, D**).

Larval habits
Larvae are found on submerged aquatic plants such as Hornwort, Milfoil, *Elodea* pondweeds and blanket weed in ponds and lakes. Where they occur they can be quite numerous.
Emergence takes place on aquatic plants that break above the water's surface and can

Figure E. Top view of larva with overall green colouration. Note that this specimen appears superficially similar to Common Blue Damselfly.
Figure F. Underside of abdomen showing fine details obscured by film of water, which creates highlights. Larvae need to be dried before examining for setae (see also Fig. **B**).

E

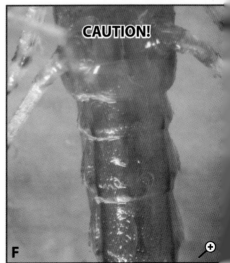

CAUTION!

F

Small Red-eyed Damselfl

occur some distance away from the pool margins.

Although currently distributed in south, east and central England this recent colonist is rapidly spreading across the UK.

Possible confusion species

This species can be confused with Common Blue, Azure, and Blue-tailed damselflies at first glance. Rows of setae on S2 and S3 will readily distinguish it (**B, H**).

Red-eyed Damselfly also have these setae but are much larger and have three distinctive dark bands rather than blotches on the caudal lamellae.

The markings on the caudal lamellae can appear superficially similar to those of Common Blue Damselfly and should be carefully examined under magnification.

Figure G. Underside of head of exuvia showing relatively narrow mask.
Figure H. Underside of abdomen of exuvia showing characteristic rows of setae on S2 and 3 (arrowed). N.B. It maybe necessary to vary the angle of light to see these clearly.
Figure I. Side view of exuvia. The dark markings on the caudal lamellae should be carefully examined. This specimen shows three dark blotches.

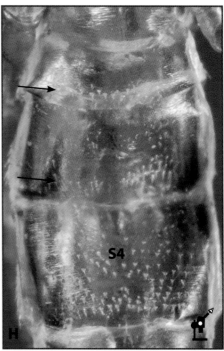

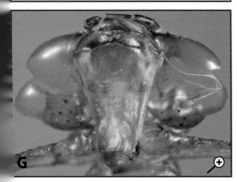

References / Suggested Reading

The following references will enable readers to delve into the available literature on identification of damselfly larvae and exuviae. Please note that these references are not cited in the text.

Andrews, S.J. (2001). Some observations on the exuviae and final instar larvae of Common Blue Damselfly *Enallagma cyathigerum* (Charpentier). *Journal of the British Dragonfly Society*, **17**(2): 35-44.

Askew, R.R. (1998) [Second edition 2004] *The Dragonflies of Europe*. Harley Books, Colchester.

Brooks, S. & Lewington, R. (2002). Key to dragonfly larvae. In: Brooks & Lewington (Eds.). *Field Guide to the Dragonflies and Damselflies of Great Britain and Ireland*. British Wildlife Publishing, Hook, England, pp.50-56.

Butler, S.G. (1995). Rearing Dragonfly larvae. *Journal of the British Dragonfly Society*, **1**(5): 74-76.

Carchini, G. (1983) A key to the italian Odonate larvae. Soc.Int.Odonatol.Rapid Comm.Suppl. 1, 1-101.

Cham, S.A. (2007). Field Guide to the larvae and exuviae of British Dragonflies. Volume 1: Dragonflies (Anisoptera). *British Dragonfly Society*. Peterborough

Crick, K. (2001). Variations in key features of the final instar larvae and exuviae of the Common Blue Damselfly *Enallagma cyathigerum* (Charpentier) *Journal of the British Dragonfly Society*, **21**(1): 27-36.

Crick, K. (2007). Observations on final instar damselfly caudal lamellae with little or no evidence of secondary tracheae. *Journal of the British Dragonfly Society*, **23**(1): 10-13.

Crick, K. (2009). Variations in key features of the final instar larvae and exuviae of the Azure Damselfly *Coenagrion puella* (Linnaeus). *Journal of the British Dragonfly Society*, **25**(1): 16-26.

Gardner, A.E. (1977). A Key to Larvae. In: Hammond, C.O. *The Dragonflies of Great Britain and Ireland*. The Curwen Press Ltd, London, pp.72-89.

Gerken, B and Sternberg, K. (1999) The exuviae of European dragonflies. Huxaria Druckerei GmbH, Hoxter. ISBN 3-9805700-4-5 - dual language German / English

Norling, U & Sahlen, G (1997) *Odonata, Dragonflies and Damselflies* in Nilsson, A. (ed) (1997) *Aquatic insects of northern Europe a taxonomic handbook* Volume 2 Odonata - Diptera. Apollo books, Stenstrup Denmark.

Seidenbusch, R. (1996) Notes on the identification of the exuviae of *Coenagrion pulchellum* (Vander Linden) and *C.puella* (Linnaeus. *Journal of the British Dragonfly Society*, **12**(1): 22-25.

Smallshire, D. & Swash, A. (2004). Identification of larvae and exuviae. In: *Britain's Dragonflies*. WildGuides, Old Basing, England, pp.148-157.

Acknowledgements

This field guide would not have been possible without the generous help of numerous individuals who offered support through advice or by providing specimens for me to photograph. I am especially grateful to Tim Beynon and Dave Smallshire for their constructive comments and advice on several early drafts and for pointing out my typos, spelling and grammar that MS Word failed to correct. To Henry Curry and Mark Tyrrell for reading final drafts with a different eye and still making constructive comments.

I would like to thank Ken Crick for sharing his ideas, larval / exuviae photographs and micrographs with me. To Graham Vick for access to his collection to photograph exuviae of *Lestes viridis and Coenagrion lunulatum*. To Jonathan Willet for providing larvae of Northern Damselfly from Scotland and to Pam Taylor for providing exuviae of Northern Damselfly. To Peter Hill for providing Variable Damselfly larvae from a population in south Wales and to Pam Taylor for larvae of the same species from Norfolk. To Peter and Kay Reeve for helping me find larvae of Scarce Blue-tailed Damselfly in Warwickshire. To the Norfolk Wildlife Trust for permission to collect larvae of Scarce Emerald Damselfly. To Ian Smith for useful advice. Mike Averill, Val Perrin and Peter Reeve for critical comments on the section on *C. pulchellum*. Graham French for the analyses of DRN data used to produce the emergence period histograms.
Special thanks go to the members of the Dragonfly Conservation Group (DCG) of the BDS for ongoing support and encouragement throughout the project.

Special acknowledgement
To Natural England; the larval photographs of Southern Damselfly were collected and photographed under a Wildlife and Countryside Act 1981 licence (Licence number: 20073583).
Larvae of this species should not be handled or collected without a licence from Natural England.

Answer to the ID challenge on page 10

The early instar larva illustrated is an Azure Damselfly *Coenagrion puella*.

Note